D1497812

Spacecraft Propulsion

craft Propulsion

D. Brown

ION SERIES
ieniecki
or-in-Chief
Institute of Technology
tterson Air Force Base, Ohio

y
Institute of Aeronautics and Astronautics, Inc.
t Promenade, SW, Washington, DC 20024-2518

American Institute of Aeronautics and Astronautics, Inc., Washington, DC

Library of Congress Cataloging-in-Publication Data

Brown, Charles D., 1930–
    Spacecraft Propulsion / Charles D. Brown.
        p.   cm.—(AIAA education series)
    Includes bibliographical references and index.
    1. Space vehicles—Propulsion systems.   I. Title.   II. Series.
TL782.B68 1995   629.47'5—dc20   95-11241
ISBN 1-56347-128-0

Second Printing

Texts Published in the AIAA Education Series

(Continued on next page.)

Published by
American Institute of Aeronautics and Astronautics, Inc., Washington, DC

Foreword

The latest text by Charles D. Brown on *Spacecraft Propulsion* in the AIAA Education Series is a companion volume to his earlier text on *Spacecraft Mission Design* published in this Series in 1992. The book includes a software disk and an appendix that serves as a user's manual that allows students to perform "hands on" computations, as if they were engaged in real life design activity. The text emphasizes the fundamentals of spacecraft propulsion design from the systems engineering viewpoint by discussing not only the propulsion systems but also the design of necessary subsystems such as propellants, pressurization, and thrust vector control. Just like his first book in the Series, this text evolved from the author's spacecraft design course at Colorado University.

Dr. Brown is eminently qualified to speak on the subject, having been involved in directing various design teams at Martin Marietta such as the Mariner 9, the first spacecraft to orbit another planet in 1971, the Viking spacecraft, and the Magellan spacecraft, which produced successful imaging of the planet Venus, and which was the first planetary spacecraft to fly on the Shuttle. In 1992, Dr. Brown received the Goddard Memorial Trophy for his Magellan project leadership and the NASA Public Service Medal, just to mention a few of his accomplishments and awards.

The AIAA Education Series embraces a broad spectrum of theory and application of different disciplines in aerospace, including aerospace design practice. More recently the Series has been expanded to encompass defense science, and technology. The Series has been in existence for over ten years and its fundamental philosophy remained unchanged: to develop texts that serve both as teaching texts for students and reference materials for practicing engineers and scientists.

J. S. Przemieniecki
Editor-in-Chief
AIAA Education Series

Preface

This book is a product of my spacecraft design course at Colorado University. Its primary purpose is to teach. To that end, the subjects are explained as clearly as possible with worked examples, so that you can teach yourself if need be. Derivations are short or referenced elsewhere. The assumptions made are those used in practice, and they are carefully covered to define the limits of usefulness of equations.

This book is different from other books in this field in three primary ways. First, the book includes PRO: AIAA Propulsion Design Software, as well as an appendix that serves as a User's Manual for the software. PRO allows you to proceed directly from understanding into professional work. Equally important, PRO provides you with the accuracy, speed, and convenience of personal computing.

Second, this book emphasizes spacecraft propulsion, which is different from launch vehicle propulsion in many ways—in dominance of monopropellants, pulsing performance, cycle life, and impulse bit repeatability, to name a few. The big launch vehicle systems are touched on, but they are not the focus.

Third, this book takes a system-level view. Rocket engines are discussed, but so are pressurization systems, propellant systems, thrust vector control systems, safe and arm systems; in short all the equipment required for a spacecraft to generate thrust is discussed. In addition, the development of system requirements is dealt with.

After deliberation, I prepared the book in the English system of units. These are the units of early propulsion development and the units in which professionals in the field still think and talk. Most propulsion reference data are still in English units. The PRO: AIAA Propulsion Design Software contains conversion routines that make it convenient to move back and forth between English and metric systems.

Charles D. Brown
February 1995

About the Author

The author has had a distinguished career as the manager of planetary spacecraft projects and as a college-level lecturer in spacecraft design. During his 30 years with Martin Marietta, he led the design team that produced propulsion systems for Mariner 9 and Viking Orbiter. Most recently he directed the team that produced the successful Venus-imaging spacecraft, Magellan. Magellan was launched on Space Shuttle Atlantis in 1989 and completed the first global map of the surface of Venus. Magellan was the first planetary spacecraft to fly on the shuttle and the first planetary launch by the United States in 10 years.

The author has instructed a popular spacecraft design course at Colorado University since 1981. This book was originally written for use in that course. He also writes software for Wren Software, Inc., a small software company he founded in 1984. He was corecipient of the Dr. Robert H. Goddard Memorial Trophy in 1992 for Magellan leadership. He has also received the Astronauts' Silver Snoopy Award in 1989; the NASA Public Service Medal, in 1992 for Magellan, and in 1976 for Viking Orbiter; and the Outstanding Engineering Achievement Award, 1989 (a team award), from the National Society of Professional Engineers, for Magellan. He is listed in *Who's Who in Science and Engineering, 1992–1993.*

Table
of
Contents

Introduction

1.1 History of Rocket Propulsion

The Chinese, Mongols, and Arabs used rockets in warfare as early as the 1200s. At the battle of Seringapatam in 1792, Indian armies used rockets against the British with such effectiveness that British soldier William Congreve designed a much improved version for British forces. The Congreve design was a 32-lb solid-fueled rocket with a range of 3000 yd. Congreve rockets were used effectively in attacks against Boulogne in 1806, Copenhagen in 1807, and the United States in 1812 (hence the phrase "rocket's red glare" in the National Anthem). In the mid-1800s, William Hale improved the Congreve rocket by introducing spin stabilization and, by the end of the Civil War, Hale's rockets had superseded Congreve's.[1] Improvements in cannon surpassed rockets and, except for use by the French in World War I, rockets were virtually extinct by the twentieth century.[2]

Konastantin Eduardovich Tsiolkowski, a Russian mathematics professor, was the first to observe that rockets were prerequisites for space exploration. As early as 1883, Tsiolkowski noted that gas expulsion could create thrust; thus, a rocket would operate even in a vacuum. In 1903, he published a milestone paper describing in detail how space flight could be accomplished with rockets. He advocated liquid oxygen and liquid hydrogen as propellants. He described staged rockets and showed mathematically that space exploration would require staging. Every time you use Eqs. (3.1)–(3.3), think of Tsiolkowski.

Dr. Robert H. Goddard, professor of physics at Clark University, was the first to design, build, and fly a liquid-propelled rocket, a feat that required more than 200 patentable inventions. Dr. Goddard, whose interest was also in space exploration, recognized that the first step into space was to devise liquid rockets that worked reliably. Could he have guessed that that first step would take his entire life? He was born in 1882, about 50 years too soon to see his dream come true. In 1919, Dr. Goddard published his milestone paper, "A Method of Reaching Extreme Altitudes." On March 26, 1926, near Worchester, Massachusetts, he flew the world's first successful liquid rocket. Propelled by liquid oxygen and gasoline, it flew for 2.5 s, reaching an altitude of 41 ft and a speed of 63 mph.

In the early 1930s, Edward Pendrey and his wife built a 7-ft liquid oxygen/gasoline rocket based on Dr. Goddard's work. It flew successfully in November 1932. The rocket was made entirely of common hardware items, at a total expenditure of $30. The following year, the Pendreys developed and fired a rocket that went 250 ft in the air and exploded.[2] Edward Pendrey is better remembered for founding the American Rocket Society, which is one of the two principal ancestors of the AIAA.

By 1929, Dr. Goddard had flown four rockets, each one better than the last. With this success and the assistance of Charles Lindbergh, he was able to gain a substantial grant from the Guggenheim Fund. With this backing, Goddard moved

Fig. 1.1 **Dr. Goddard with Lindbergh and Guggenheim. (Courtesy Permanent Collection, Roswell Museum and Art Center, Roswell, New Mexico. Photograph by Esther C. Goddard.)**

his work to the isolation of the desert near Roswell, New Mexico. Dr. Goddard is shown with Charles A. Lindbergh and Harry F. Guggenheim in Fig. 1.1; the photograph was taken in front of his launch tower near Roswell, by his wife Ester. Here he developed pump-fed stages, clustered stages, and pressurization systems; worked on gyro control systems; and developed the idea of sounding rockets. He worked his rockets up to a size of 500 lb, altitudes of 1.2 miles, and speeds of 700 mph. When Goddard died in 1945, he had developed the first version of every piece of equipment that would be needed for a vehicle like Saturn V. His work would not be fully appreciated for another 10 years—at least, not in the United States.

A test data sheet from one of his tests is shown in Fig. 1.2. Note that he had developed turbine driven pumps by 1941 when this test was run. Also note that he had discovered the concept of specific impulse which he called "average thrust per lb. total flow per second."

Enthusiasm for rocketry was more vigorous in Russia and Germany, and Dr. Goddard's work was followed more carefully there than at home. In Russia, the government funded research as early as 1930 through two research groups, the Leningrad Gas Dynamics Laboratory (the GDI) and the Group for the Study of Jet Propulsion (the GIRD).[2] It was the latter group, under the leadership of Sergei Korolyev, that was effective in producing the series of Russian successes in propulsion that led to the launch of Sputnik.

In Germany, the government instituted vigorous research in rocketry in order to circumvent the armament restrictions that resulted from World War I. The work was carried out under the sponsorship of Karl Becker, head of ballistics and armament for the German army. He appointed Walter Dornberger to develop the rockets. Dornberger was joined by a young enthusiast, Werner von Braun. In 1932, Dornberger led the construction of the test site and von Braun produced and successfully fired their first liquid rocket. The group had started their work in a

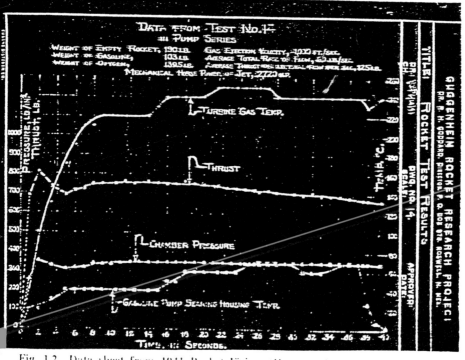

Fig. 1.2 Data sheet from 1941 Rocket Firing. (Courtesy Permanent Collection, Roswell Museum and Art Center, Roswell, New Mexico.)

suburb of Berlin that clearly was not suitable for continued firings. In 1935, they moved to an isolated peninsula near the Baltic Sea called Peenemünde. It was at Peenemünde that the second von Braun design, the A-2, was built and successfully launched, followed by the A-3 and, in 1942, the A-4. The A-4 was renamed the Vengeance-2, or V-2, by the army and placed in production. The V-2, a remarkable vehicle for 1942, was 46 ft high, was single-stage, weighed 25,000 lb, and was capable of carrying a 2000-lb payload 118 miles. It was the first vehicle of any kind to break the sound barrier. The propellants were liquid oxygen and alcohol. (The fuel was changed from the gasoline that Goddard used to alcohol because the Germans had more potatoes than petroleum.) Six thousand V-2s were made; 2000 fell on England, too late to influence the course of the war.

In 1945, it was clear to the team at Peenemünde that the war was lost. With their facility in the path of the Russian army, von Braun and over 100 members of his staff decided that the United States was a preferable captor. In an incredible move, they left Peenemünde and moved tons of equipment and documents to the Hartz mountains, where they were certain to be captured by U.S. troops.[2] Von Braun and his staff were moved to Redstone Arsenal in Huntsville, Alabama, which later became Marshal Space Flight Center. Von Braun's team became the nucleus from which the U.S. rocket capability grew. Von Braun was a strong and innovative leader in the development of propulsion for the United States.

1.2 Propulsion System Types

Current propulsion technology provides four basically different propulsion choices: cold-gas systems, monopropellant systems, bipropellant systems, and solid motor systems. Each has its niche in spacecraft design. The selection of propulsion system type has substantial impact on the total spacecraft and is a key selection in early design. Several other system types are feasible, ion propulsion and nuclear propulsion to name two, but flight use of these systems is minimal.

Cold gas systems. Almost all of the spacecraft of the 1960s used a cold-gas system (see Fig. 1.3). It is the simplest choice and the least expensive. Cold-gas systems can provide multiple restarts and pulsing. The major disadvantage of the system is low specific impulse (about 50 s) and low thrust levels (less than a pound), with resultant high weight for all but the low total impulse missions.

Monopropellant Systems. The next step up in complexity and cost is the monopropellant system (Fig 1.4). It can supply pulsing or steady-state thrust. The specific impulse is about 225 s, and thrusts from 0.1 lb to several hundred pounds

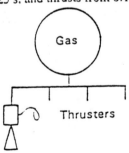

Fig. 1.3 Cold-gas system.

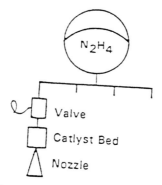

Fig. 1.4   Monopropellant system.

are available. The system is the common choice for attitude control and midrange impulse requirements. The only monopropellant in flight use is hydrazine.

Bipropellant systems.   Although they rank highest on the scale of complexity and expense. bipropellant systems (Fig 1.5) provide a very versatile and high performance. with a specific impulse I_{sp} of about 310 s and a wide range of thrust capability (from a few pounds to many thousands of pounds). Bipropellant systems can be used in pulsing or steady-state modes. The most common propellants for spacecraft use are nitrogen tetroxide and monomethyl hydrazine. Bipropellant systems are not as common as monopropellants because the high total impulse with restart is not usually required.

Figure 1.4 shows the system weights for propulsion systems vs the total impulse delivered.[3] These systems are used in pulse mode for attitude control; hence, solid systems are not shown. Note that bipropellant systems are actually heavier than monopropellant systems at impulses below about 10,000 lb-s. At low impulses, the additional hardware required for a bipropellant system outweighs the performance gained with high specific impulse. Below total impulse levels of about 100,000 lb-s, monopropellants should be considered.

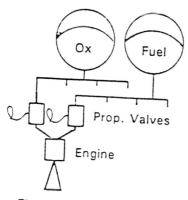

Fig. 1.5   Bipropellant system.

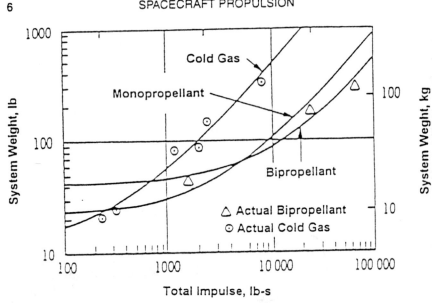

Fig. 1.6   System weight comparison (from Ref. 3, p. 343).

Solid motor systems.   When all of the impulse is to be delivered in a single burn and the impulse can be accurately calculated in advance (shut down/restart is not current state of the art for solids), solid motor systems (Fig 1.7) are candidates. Examples of situations of this type are planetary orbit insertion and geosynchronous apogee burns (apogee kick motors). If the single burn criterion is met, solids provide simplicity, reasonable performance (specific impulse about 290 s), and costs comparable to monopropellant systems.

The characteristics of each system type are summarized in Table 1.1. The impulse range shown is the range over which the system type is commonly used; it does not indicate a technical or feasibility limit for the system type. The complexity rankings consider component count for complete systems. The choices for most and least complex are not likely to cause debate; comparing solid and monopropellant systems is not so clear and would require a trade study of specific applications.

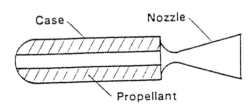

Fig. 1.7   Solid motor.

Table 1.1   Characteristics of propulsion system types

	Cold gas	Monoprop	Biprop	Solid
Specific impulse, s	50	225	310	290
Thrust range, lb	0.01–0.02	0.1–600	>2	>300
Impulse range, lb-s	$<10^3$	10^3-10^5	10^4-10^8	10^4-10^8
Min impulse bit, lb-s	0.0002	0.003	0.03	N
Complexity	Least	Midrange	Most	Midrange
S/C contamination	N	N	Y	Y
Restart	Y	Y	Y	N
Pulsing	Y	Y	Y	N
Throttling	N	Y	Y	N

1.3   PRO: AIAA Propulsion Design Software

This book includes PRO, the AIAA Propulsion Design Software. With PRO, you can design complete systems with the accuracy, speed, and convenience of personal computing. It no longer requires a mainframe to do quality work in this field. The PRO manual is Appendix A in this volume.

1.4   Arrangement of the Book

Chapter 2 deals with the basic equations of propulsion design. Chapter 3 describes how the performance requirements of a propulsion system are derived. Chapters 4–7 take up the considerations peculiar to each of the four basic types of spacecraft propulsion systems. Monopropellant systems are described first since they are most common. Equipment common to each system, such as tankage and pressurization, is discussed in Chapter 4 and referenced thereafter.

The Appendices were designed to provide the working professional with ready reference material. The manual for the AIAA Propulsion Design Software is Appendix A. Equations, design data, and conversion factors are in Appendix B. Appendix C is a glossary of propulsion terms.

2
Theoretical Rocket Performance

A rocket generates thrust by accelerating a high-pressure gas to supersonic velocities in a converging-diverging nozzle. In most cases, the high-pressure gas is generated by high-temperature decomposition of propellants. As shown in Fig. 2.1, a rocket consists of a chamber, throat, and nozzle.

In a bipropellant rocket engine, the gases are generated by the rapid combustion of liquid propellants in the combustion chamber. A liquid fuel and a liquid oxidizer, such as liquid hydrogen and liquid oxygen, are used. In a monopropellant system, only one propellant is used. High-pressure, high-temperature gases are generated by decomposition. Hydrazine is the most common monopropellant. In a solid system, the fuel and oxidizer are mechanically mixed and cast as a propellant grain. The grain occupies most of the volume of the combustion chamber. In a cold-gas system, no combustion is involved. A gas—for example helium—is stored at high pressure and injected into the chamber without combustion.

2.1 Thrust

Rocket thrust is generated by momentum exchange between the exhaust and the vehicle and by the pressure imbalance at the nozzle exit. The thrust due to momentum exchange can be derived from Newton's second law,

$$F_m = \dot{m} \Delta V \tag{2.1}$$

$$F_m = \dot{m}(V_e - V_0) \tag{2.2}$$

where

F_m = thrust generated due to momentum exchange, lb
$\dot{m}$ = mass flow rate of the propellant or mass flow rate of the exhaust gas, slugs/s
V_e = average velocity of the exhaust gas, fps
V_0 = initial velocity of the gases, fps

Since the initial velocity of the gases is zero,

$$F_m = \dot{m} V_e \tag{2.3}$$

$$F_m = \frac{\dot{w}_p}{g_c} V_e \tag{2.4}$$

where $\dot{w}_p$ = weight flow rate of propellants flowing into the chamber, lb/s.

Equations (2.3) and (2.4) assume that the velocity of the exhaust gas acts along the nozzle centerline. Note that g_c is the gravitational constant used to convert weight to mass. The gravitational constant is equal to 32.1741 fps^2 and is invariant

9

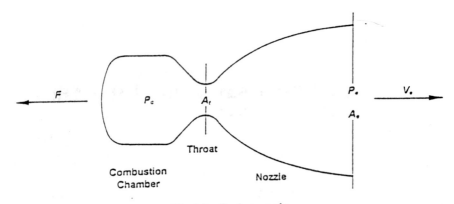

Fig. 2.1 Rocket nozzle.

with distance from the Earth's center. It is not to be confused with the acceleration of gravity, which varies inversely with the square of the distance to the center of the central body.

In addition to the thrust due to momentum, thrust is generated by a pressure-area term at the nozzle exit. If the nozzle were exhausting into a vacuum, the pressure-area thrust would be

$$F_p = p_e A_e \qquad (2.5)$$

If the ambient pressure is not zero,

$$F_p = P_e A_e - P_a A_e \qquad (2.6)$$
$$F_p = (P_e - P_a) A_e \qquad (2.7)$$

where

F_p = thrust due to exit plane pressure, lb
P_e = static pressure in the exhaust gas, psia
A_e = area of the nozzle exit when hot, in.^{2}
P_a = ambient static pressure, psia

Because the total thrust on the vehicle is the sum of the thrust due to momentum and thrust due to exit plane pressure, the fundamental relationship for thrust becomes

$$F_m = \frac{\dot{w}_p}{g_c} V_e + (P_e - P_a) A_e \qquad (2.8)$$

For any given engine, thrust is greater in a vacuum than at sea level by an amount equal to $P_a A_e$. A rocket is the only engine that will operate in a vacuum, and it is substantially more efficient in a vacuum, as shown in Fig. 2.2.

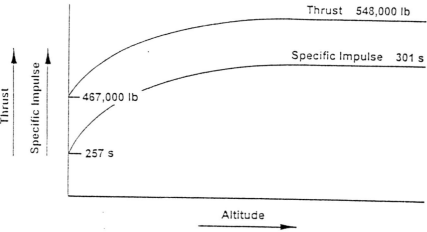

Fig. 2.2   Thrust vs altitude; sea level and altitude performance of the Titan IV, stage I, engine.

Converging-Diverging Nozzle Performance

A rocket engine takes advantage of the supersonic velocity increases that can be achieved in converging-diverging nozzles. Figure 2.3 shows the pressure in a converging nozzle as a function of axial position in the nozzle. Line A in Fig. 2.3 shows the variation in gas pressure in a converging-diverging nozzle during subsonic flow. The pressure decreases and velocity increases as the gas approaches the throat. The velocity reaches a maximum at the throat. If the velocity is subsonic, the velocity decreases in the diverging section and the pressure increases, reaching ambient pressure as velocity returns to zero. A quite different result occurs if the gas velocity is sonic at the throat, as at line B in Fig. 2.3. In the sonic case, the velocity continues to increase in the divergent section and pressure continues to decrease. If the nozzle is overexpanded (exit plane pressure below ambient pressure), the flow separates from the wall, causing a sharp pressure rise; then, recirculation and significant performance losses occur. If the exit plane pressure just matches ambient pressure, as at line C in Fig. 2.3, separation does not occur and performance is maximum. This is called *optimum expansion.*

For engines that must operate in atmosphere, for example, launch vehicle engines, the expansion of the nozzle is designed so that $P_e = P_a$ at design altitude. Figure 2.4 shows thrust vs altitude for engines optimized at sea level and at 40,000 ft and for a rubber engine optimized at all altitudes.

For optimum expansion, the pressure thrust term is zero and may be dropped. Thus,

$$F_m = \frac{\dot{w}_p}{g_c} V_c \qquad (2.9)$$

High-area-ratio spacecraft engines have a low exit pressure and are very nearly at optimum expansion in a vacuum. It is common practice to adjust velocity to compensate for the small pressure-area thrust produced by a real vacuum engine.

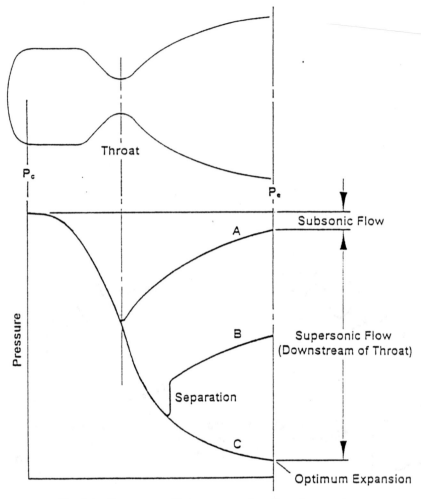

Fig. 2.3   Pressure profile in a converging-diverging nozzle.

The adjusted velocity is called the *effective exhaust velocity*. Effective exhaust velocity is readily determined on a test stand by measuring thrust and propellant flow rate. This common practice is more accurate than attempting to determine the actual exhaust velocity profile.

2.2   Ideal Rocket Thermodynamics

If we idealize the flow in a rocket engine, we can use thermodynamics to predict rocket performance parameters to within a few percent of measured values. The following seven assumptions define what is known as *theoretical performance*:[5]

1) The propellant gases are homogeneous and invariant in composition throughout the nozzle. This condition requires good mixing and rapid completion of combustion and, in the case of solids, homogeneous grain. Good design will provide these conditions.

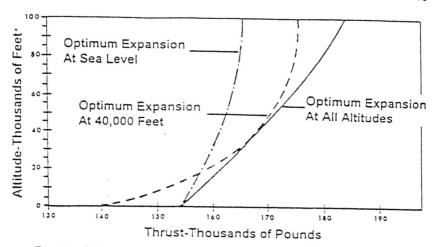

Fig. 2.4   Optimum expansion for various altitudes (from Ref. 4, pp. 3–5).

2) The propellant gases follow the perfect gas laws. The high temperature of rocket exhaust is above vapor conditions, and these gases approach perfect gas behavior.

3) There is no friction at the nozzle walls and, therefore, no boundary layer.

4) There is no heat transfer across the nozzle wall.

5) Flow is steady and constant.

6) All gases leave the engine axially.

7) The gas velocity is uniform across any section normal to the nozzle axis.

Assumptions 3, 4, 6, and 7 permit the use of one-dimensional isentropic expansion relations. Assumption 1 defines what is called *frozen equilibrium* conditions. The gas composition can be allowed to vary from section to section in what is called a *shifting equilibrium* calculation. Shifting equilibrium calculation accounts for exothermic recombinations that occur in the exhaust stream; higher propellant performance results. With these assumptions, the performance parameters can be calculated as well as measured and the results compared to yield efficiencies.

Conditions at Any Nozzle Section

Ideal flow conditions in any section of a converging-diverging nozzle can be shown to be[5–8]

$$\frac{V_x}{V_t} = \sqrt{\frac{k+1}{k-1}\left[1 - \left(\frac{2}{k+1}\right)\left(\frac{p_x}{p_t}\right)^{\frac{k-1}{k}}\right]} \qquad (2.10)$$

$$\frac{T_x}{T_t} = \left(\frac{p_x}{p_t}\right)^{\frac{k-1}{k}} \qquad (2.11)$$

$$\frac{A_x}{A_t} = \frac{V_t}{V_x}\left(\frac{p_t}{p_x}\right)^{\frac{1}{k}} \qquad (2.12)$$

where

A = area of the nozzle cross section, ft^2
k = ratio of specific heats of the gas, C_p/C_v
p = absolute pressure of the gas, psia
T = absolute total temperature of the gas, °R
V = velocity of the gas, fps

Subscript t denotes conditions at the throat, and x denotes conditions at any other section taken perpendicular to the nozzle centerline.

Critical Pressure Ratio

The critical pressure ratio is the pressure ratio required for sonic, or choked, flow at the throat. Equation (2.10) can be reduced to an expression for critical pressure by setting chamber velocity equal to zero.

$$\frac{P_c}{P_t} = \left(\frac{k+1}{2}\right)^{\frac{k}{k-1}} \qquad (2.13)$$

Equation (2.13) shows that critical pressure ratio is a pure function of specific heat ratio. The range of specific heat ratios encountered in propellant gases is narrow; the lowest is about 1.2, and the maximum possible (helium) is 1.67. The corresponding critical pressure ratios are 1.77 and 2.12. For an engine operating at sea level, a chamber pressure of about 35 psia is adequate to assure supersonic flow in the nozzle. The higher the altitude, the lower the chamber pressure for sonic flow. Because the chamber pressures of real engines are usually greater than 100 psia, it is safe to assume sonic flow at the throat of a rocket nozzle.

From thermodynamics, it can be shown that the sonic velocity is

$$a = \sqrt{kg_c RT} \qquad (2.14)$$

where

R = specific gas constant, $1545/M$, ft-lbf/lbm·°R
M = molecular weight of gas
T = gas absolute total temperature, °R
a = sonic velocity, fps

Thus, the velocity at the throat of a rocket nozzle is

$$V_t = \sqrt{kg_c RT_c} \qquad (2.15)$$

Because the gas is at rest in the chamber, the gas temperature in the chamber, T_c, is the total temperature.

Gas Velocity

If it is assumed that the throat velocity is sonic, Eq. (2.10) may be rearranged to produce an expression for velocity at any plane in the expansion section of the nozzle:

$$V_x = \sqrt{\frac{2kg_cRT_c}{k-1}\left[1 - \left(\frac{P_x}{P_c}\right)^{\frac{k-1}{k}}\right]} \qquad (2.16)$$

where

V_x = velocity of the gas at any plane in a nozzle, fps
P_x = pressure at the reference plane, psia

Velocity at the exit plane is a special case of Eq. (2.16), where $V_e = V_x$, $P_e = P_x$, and

$$V_e = \sqrt{\frac{2kg_cRT_c}{k-1}\left[1 - \left(\frac{P_e}{P_c}\right)^{\frac{k-1}{k}}\right]} \qquad (2.17)$$

Equation (2.17) is for optimum expansion. Zucrow[8] refers to results from Eq. (2.17) as ideal exhaust velocity to differentiate it from the effective exhaust velocity.

The maximum exhaust velocity occurs when pressure ratio P_e/P_c is zero,

$$(V_e)_{max} = \sqrt{\frac{2kg_cRT_c}{k-1}} \qquad (2.18)$$

Equations (2.17) and (2.18) show that exit velocity, and hence thrust, increases as chamber temperature and gas constant increase. The pressure ratio and specific heat ratio are minor influences. It is convenient to remember that thrust and exit velocity are proportional to $\sqrt{T_c/M}$. Table 2.1 shows the correlation of T_c/M and V_e for some rocket propellant combinations.[6] Exhaust velocities in Table 2.1 were calculated for vacuum conditions, optimum expansion, and an area ratio of 30.

Table 2.1 Characteristics of some propellants

Propellants	T_c, °R	M	T_c/M	V_e, fps
Oxygen/hydrogen	6395	10.0	25.3	14,520
Oxygen/RP-1[a]	6610	23.3	16.8	11,370
N_2O_4/MMH[b]	6100	21.5	16.8	10,820
Monopropellant N_2H_4	2210	13	13.0	7,630
He (cold gas)	520	4	11.4	5,090
N_2 (cold gas)	520	28	3.4	2,190

[a]RP-1 (rocket propellant-1) is a common hydrocarbon fuel similar to kerosene.
[b]MMH is monomethyl hydrazine.

Specific Impulse, I_{sp}

Specific impulse is the premier measurement of rocket performance. It is defined as the thrust per unit weight flow rate of propellant,

$$F = I_{sp}\dot{w}_p \tag{2.19}$$

The proper units for I_{sp} are lbf-s/lbm but, in practice, this cumbersome unit is abbreviated to seconds (s). Specific impulse, like thrust, is higher in a vacuum than at sea level. I_{sp} is the conventional method of comparing propellants, propellant combinations, and the efficiency of rocket engines.

Comparing Eqs. (2.9) and (2.19) shows that I_{sp} can also be expressed as,

$$I_{sp} = V_e/g_c \tag{2.20}$$

where V_e = effective exhaust gas velocity in feet per second.

By assuming that effective and ideal exhaust velocities are equal, that is, that optimum expansion is achieved, we can obtain a thermodynamic relationship for I_{sp} from Eq. (2.20),

$$I_{sp} = \sqrt{\frac{2kRT_c}{g_c(k-1)}\left[1 - \left(\frac{P_e}{P_c}\right)^{\frac{k-1}{k}}\right]} \tag{2.21}$$

Equation (2.21) shows that I_{sp} is a thermodynamic property of the propellants; for a given expansion ratio, it is independent of motor design. Like V_e, I_{sp} is proportional to $\sqrt{T_c/M}$. In a vacuum, Eq. (2.21) reduces to

$$I_{sp} = \sqrt{\frac{2kRT_c}{g_c(k-1)}} \tag{2.22}$$

Equation (2.22) shows that, in a vacuum, specific impulse is independent of chamber pressure; therefore, spacecraft engines can (and do) have low chamber pressures compared to their sea level counterparts. Monopropellants are an exception to this rule. For monopropellants, a second-order chamber pressure dependence results from the nonideal dissociation of ammonia, which will be discussed later.

Table 2.2 compares the theoretical, vacuum, shifting equilibrium I_{sp} of propellants at an area ratio of 30.[5]

To specify specific impulse fully, it is necessary to state:
1) Chamber pressure.
2) Area ratio.
3) Ambient pressure.
4) Shifting or frozen equilibrium conditions
5) Real or theoretical.

Specific impulse calculated from Eq. (2.21) is theoretical, frozen, and optimum expansion; chamber pressure and altitude can be anything you wish. If gas composition is allowed to vary in the nozzle, shifting equilibrium conditions prevail, and a slightly higher I_{sp} results. Figure 2.5 compares shifting and frozen theoretical specific impulse for several propellant combinations. Specific impulse from real

Table 2.2 Theoretical I_{sp} for some propellants

Oxidizer	Fuel	MR^a	T_c, °F	$C^{*,b}$ fps	I_{sp}, s	M	k
Fluorine	Hydrazine	2.30	7930	7280	422	19.4	1.33
	Hydrogen	7.60	6505	8355	470	11.8	1.33
Nitrogen Tetroxide	Hydrazine	1.00	5450	5830	336	18.9	1.26
	MMH	2.19	5640	5710	336	21.5	1.24
	UDMH^c	2.60	5685	5650	333	21.0	1.25
Oxygen	Hydrogen	4.00	5935	7985	451	10.0	1.26
	UDMH	1.67	6045	6115	363	21.3	1.25
	RP-1	2.55	6145	5915	353	23.3	1.24
Monoprop	Hydrazine	—	1750	4280	237	13	1.27

^{a}Mixture ratio. ^{b}Characteristic velocity. ^{c}Unsymmetrical dimethylhydrazine.

engines lies between shifting and frozen calculations. In practice, about 93% of theoretical I_{sp} is achieved during steady-state operation.

Total Impulse, I

Impulse is defined as a change in momentum caused by a force acting over time,

$$I = Ft \tag{2.23}$$

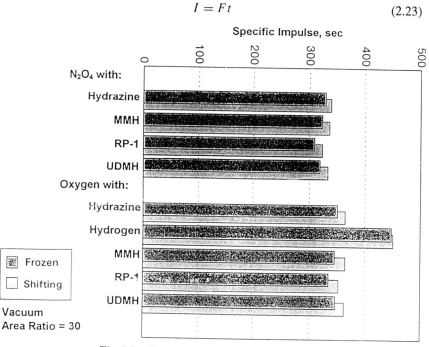

Fig. 2.5 Frozen and shifting theoretical I_{sp}.

where

I = impulse delivered to the spacecraft, lb-s
F = thrust, lb
t = time, s

For cases in which thrust is not constant, as in solid rockets, impulse is the area under the thrust-time curve. Impulse, sometimes called *total impulse*, is related to specific impulse as follows:

$$I = W_p I_{\text{sp}} \qquad (2.24)$$

where W_p is the total weight of propellant consumed.

Propulsion systems are rated based on impulse; this is a particularly useful measure of solid rocket motor systems although it is used for all types.

Weight Flow Rate

The weight flow rate of gas through a supersonic isentropic nozzle is[6-8]

$$w_p = \frac{P_c A_t k g_c}{\sqrt{k g_c R T_c}} \sqrt{\left(\frac{2}{k+1}\right)^{\frac{k+1}{k-1}}} \qquad (2.25)$$

Equation (2.25) shows that, for a given propellant and stagnation temperature, the flow rate through a nozzle is proportional to chamber pressure and throat area and those parameters only.

Area Ratio

The area ratio of a rocket engine is the ratio of the exit area to the throat area,

$$e = \frac{A_e}{A_t} \qquad (2.26)$$

where

e = area ratio, unitless
A_e = exit area measured hot
A_t = throat area measured hot

Area ratio is a measure of the gas expansion provided by an engine. Optimum area ratio provides an exit plane pressure equal to local ambient pressure. For a sea level or first-stage engine, the optimum area ratio is selected near midpoint of the flight; an area ratio of about 10 is common. For a spacecraft engine (or an upper-stage engine), the optimum area ratio is infinite; the largest area ratio allowed by space and weight is used. An engine with an area ratio less than optimum is referred to as underexpanded. The exhaust gas in an underexpanded nozzle can separate from the wall, causing recirculation and a substantial loss of efficiency.

Area ratio and pressure ratio are related as follows:[6]

$$\frac{A_e}{A_t} = \frac{\sqrt{k\left(\frac{2}{k+1}\right)^{\frac{k+1}{k-1}}}}{\left(\frac{P_e}{P_c}\right)^{\frac{1}{k}}\sqrt{\frac{2k}{k-1}\left[1-\left(\frac{P_e}{P_c}\right)^{\frac{k-1}{k}}\right]}} \qquad (2.27)$$

Figure 2.6 shows pressure ratio as a function of area ratio for various specific heat ratios and optimum expansion.

Characteristic Velocity, C*

Characteristic velocity C^* is an empirical rocket parameter used to separate the thermochemical performance of propellants from the performance of a particular engine. Characteristic velocity is expressed as follows:

$$C^* = \frac{P_c A_t g_c}{w_p} \qquad (2.28)$$

Characteristic velocity measures combustion performance by indicating how many pounds of propellant must be burned to maintain chamber pressure. It is independent of the performance of the nozzle and does not vary with ambient pressure. The theoretical relation for C^* is[6]

$$C^* = \frac{\sqrt{kg_c R T_c}}{k\sqrt{\left(\frac{2}{k+1}\right)^{\frac{k+1}{k-1}}}} \qquad (2.29)$$

Equation (2.29) shows that C^* is a pure function of k, R, and T_c; therefore, C^* is a propellant property. Like exhaust velocity and specific impulse, characteristic velocity is proportional to $\sqrt{T_c/M}$. About 95% of theoretical C^* is achieved in practice during steady-state performance.

Thrust Coefficient, C_f

Thrust coefficient is a useful term that first arose during rocket engine testing. It is the proportional constant between thrust and the product of chamber pressure and throat area.

$$F = P_c A_t C_f \qquad (2.30)$$

where

P_c = chamber pressure, psia
A_t = area of the nozzle throat when hot, in.^{2}

The improvement in thrust provided by the nozzle is characterized by thrust coefficient; an engine without a diverging section would have a thrust coefficient

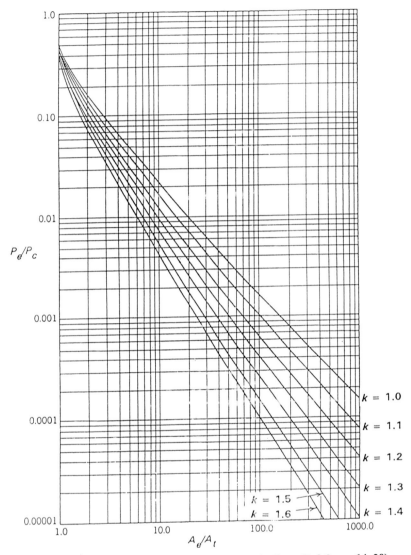

Fig. 2.6   **Pressure ratio as a function of area ratio (from Ref. 9, pp. 14–20).**

of one. Thrust coefficient is unitless and larger than one. Thrust coefficient, like characteristic velocity, aids the separation of engine performance into parts. The two are related as follows:

$$C = \frac{C_f}{C^*} \tag{2.31}$$

$$I_{sp} = \frac{C_f C^*}{g_c} \tag{2.32}$$

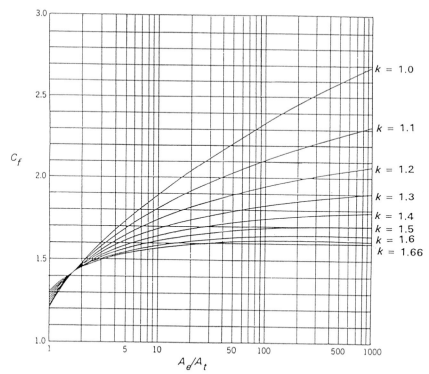

Fig. 2.7 Theoretical C_f vs area ratio for vacuum conditions (from Ref. 9, pp. 14–22).

From thermodynamics, theoretical C_f is[6]

$$C_f = \sqrt{\frac{2k^2}{k-1}\left(\frac{2}{k+1}\right)^{\frac{k+1}{k-1}}\left[1-\left(\frac{P_e}{P_c}\right)^{\frac{k-1}{k}}\right]} + \left(\frac{P_e-P_a}{P_c}\right)\frac{A_e}{A_t} \qquad (2.33)$$

Figure 2.7 shows theoretical C_f as a function of k and area ratio for vacuum conditions. In practice, about 98% of theoretical C_f is achieved in steady-state performance.

Mixture Ratio, MR

Mixture ratio is an important parameter for bipropellant systems. It is the ratio of oxidizer to fuel flow rate on a weight basis:

$$MR = \frac{w_o}{w_f} \qquad (2.34)$$

where

w_o = oxidizer weight flow rate, lb/s
w_f = fuel weight flow rate, lb/s

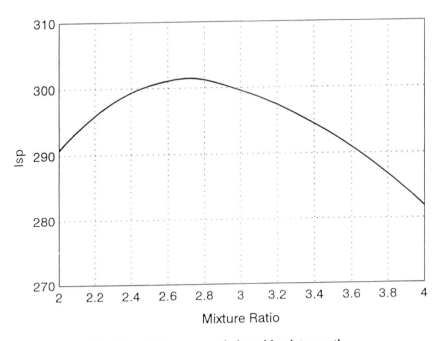

Fig. 2.8   Performance variation with mixture ratio.

Each propellant combination has an optimum mixture ratio that produces maximum I_{sp}. At optimum mixture ratio, the proportions of oxidizer and fuel react to produce maximum energy release. Figure 2.8 shows the variation in vacuum I_{sp} as a function of mixture ratio for liquid oxygen and RP-1. Optimum mixture ratios for some bipropellant combinations are shown in Table 2.2.[6]

During the manufacture of an engine, the propellant flow passages are orificed to control the mixture ratio consumed by the thrust chamber. The mixture ratio of an engine may be different from that of the thrust chamber because of propellant used by accessories. The Titan IV, stage I, engine, for example, has a nominal mixture ratio of 1.91; the thrust chamber has a mixture ratio of 2.00.

Volumetric mixture ratio is sometimes used in conjunction with tank sizing; mixture ratio can be converted to volumetric mixture ratio using Eq. (2.35). (By convention, mixture ratio is always by weight unless otherwise specified.)

$$VMR = MR\frac{\rho_f}{\rho_o} \qquad (2.35)$$

where

ρ_f = density of the fuel, lb/ft^3
ρ_o = density of the oxidizer, lb/ft^3

Table 2.3 Bulk mixture ratio of some propellant combinations

Propellant combination	Bulk density g/cc	@ Mixture ratio
N_2O_4/MMH	1.17	1.75
N_2O_4/N_2H_4	1.25	1.20
Oxygen/hydrogen	0.28	4.0
Oxygen/RP-1	1.01	2.20

The volumetric mixture ratio consumed by the engine has a major effect on system design because it determines the relative sizes of the propellant tanks.

Bulk Density

It is often convenient to use the bulk density of a propellant combination to expedite calculation. Bulk density is the mass of a unit volume of a propellant combination "mixed" at the appropriate mixture ratio. Bulk density may be computed as follows:

$$\rho_b = \frac{MR + 1}{\dfrac{MR}{\rho_o} + \dfrac{1}{\rho_f}} \tag{2.36}$$

where ρ_b = bulk density of a propellant combination in pounds per cubic foot. The bulk density and mixture ratio of some propellant combinations are shown in Table 2.3.

Problems

2.1 A man is sitting in a rowboat throwing bricks out the back. Each brick weighs 5 lb; the man is throwing them at a rate of six bricks per min with a velocity of 32 fps. What is his thrust and I_{sp}?

2.2 If a propulsion system delivers 1000 lb of thrust with a propellant flow rate of 3.25 lb/s, what is the specific impulse?

2.3 The sea level characteristics of a bipropellant engine are:

$$F = 95,800 \text{ lb}$$
$$I_{sp} = 235 \text{ s}$$
$$\text{Throat area} = 248 \text{ in.}^2$$
$$\text{Area ratio} = 8$$

What is the vacuum thrust and specific impulse?

2.4   The characteristics of an engine are:

$$F = 18.67 \text{ lb}$$
$$P_c = 97 \text{ psia}$$
$$A_t = 0.118 \text{ in.}^2$$
$$I_{sp} = 289 \text{ s}$$
$$\text{Area ratio} = 40$$

The thrust and I_{sp} were measured in a test chamber with an ambient pressure of 0.282 psia. What vacuum thrust and I_{sp} would you expect of this engine? What is the vacuum thrust coefficient? Is this a monopropellant or bipropellant engine?

2.5   An engine delivers an exhaust velocity of 9982 fps in a vacuum. What is the vacuum I_{sp} of this engine?

2.6   An engine has the following characteristics:

$$P_c = 500 \text{ psia}$$
$$A_t = 365 \text{ in.}^2$$
$$\text{Ratio of specific heats} = 1.2 \text{ (exhaust gas)}$$
$$\text{Area ratio} = 30$$

What is the vacuum thrust coefficient and vacuum thrust of this engine?

2.7   An engine operating with an area ratio of 50 on monopropellant hydrazine in a vacuum delivers an exhaust gas with a chamber temperature of 1850°F, a ratio of specific heats of 1.27, a molecular weight of 14, and a pressure ratio P_c/P_e of 845. What is the theoretical I_{sp}, C^*, C_f, and exhaust velocity of this gas? What actual specific impulse would you expect from this engine?

2.8   If an engine has a thrust coefficient of 1.75 and a characteristic velocity of 7900 fps, what is the specific impulse?

2.9   The mixture ratio for an engine is 2.2. What is the volumetric mixture ratio at 68°F if the oxidizer is nitrogen tetroxide and the fuel monomethyl hydrazine?

2.10   A rocket engine has a throat area of 0.768 in.^{2}, a chamber pressure of 700 psi, a steady-state propellant flow rate of 3.10 lb/s, and a vacuum thrust coefficient of 1.81. What is the steady-state vacuum specific impulse of this engine? What is the steady-state vacuum thrust? Is this a monopropellant, bipropellant, cold-gas, or solid motor?

3
Propulsion Requirements

Before we discuss how thrust and impulse are generated, let us consider why they are needed and how much is needed. The two major sources of requirements are the design of the mission and the attitude-control system. Table 3.1 shows some of the tasks performed by spacecraft propulsion.

3.1 Mission Design

Mission design analysis is covered thoroughly in the text *Spacecraft Mission Design*[10] and elsewhere. One of the main products of mission design analysis is a statement of the ΔV required for the maneuvers listed in Table 3.1. In order to convert these velocity change requirements to propellant requirements, we need the Tsiolkowski equation and its corollaries:

$$\Delta V = g_c I_{\text{sp}} \ln\left(\frac{W_i}{W_f}\right) \tag{3.1}$$

$$W_p = W_i \left[1 - \exp\left(-\frac{\Delta V}{g_c I_{\text{sp}}}\right)\right] \tag{3.2}$$

$$W_p = W_f \left[\exp\left(\frac{\Delta V}{g_c I_{\text{sp}}}\right) - 1\right] \tag{3.3}$$

where,

W_i = initial vehicle weight, lb
W_f = final vehicle weight, lb
W_p = propellant weight required to produce the given ΔV, lb
ΔV = velocity increase of the vehicle, fps
g_c = gravitational constant, 32.1740 ft/s^2

Equations (3.2) and (3.3) yield the propellant weight required for a spacecraft of a given weight to perform a maneuver. It is important not to use Eqs. (3.1)–(3.3) in cases where drag or unbalanced gravity forces are significant.

Finite Burn Losses

Mission design calculations assume that velocity is changed at a point on the trajectory, that is, that the velocity change is instantaneous. If this assumption is not valid, serious energy losses can occur. These losses, called *finite burn losses*, are caused by the rotation of the spacecraft velocity vector during the burn, as shown in Fig. 3.1.

Table 3.1   Spacecraft propulsion functions

Task	Description
Mission design	(Translational velocity change)
Orbit changes	Convert one orbit to another
Plane changes	
Orbit trim	Remove launch vehicle errors
Stationkeeping	Maintain constellation position
Repositioning	Change constellation position
Attitude Control	(Rotational velocity change)
Thrust vector control	Remove vector errors
Attitude control	Maintain an attitude
Attitude changes	Change attitudes
Reaction wheel unloading	Remove stored momentum
Maneuvering	Repositioning the spacecraft axes

In Fig. 3.1, it is assumed that the thrust vector is held inertially fixed during the burn. It oversimplifies the case in two respects:

1) The orbit at the end of the burn is not the same orbit as at the start. The orbit may not even be the same type; the initial orbit might be a circle and the final orbit an ellipse. The orbital elements are continually changing during the burn.

2) If the final orbit in Fig. 3.1 were an ellipse, θ would be increased by the amount of the final flight-path angle. In reality, there will be a continuous change in flight-path angle as a function of time.

Also, note that the change in spacecraft velocity is not linear with time [see Eq. (3.1)]; acceleration is much larger at the end of the burn because the spacecraft is lighter.

The finite burn losses can be a significant percentage of the maneuver energy. Figure 3.2 shows the finite burn losses for a Venus orbit insertion as a function of thrust-to-weight ratio for a specific aircraft.[10]

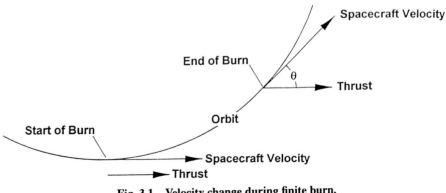

Fig. 3.1   Velocity change during finite burn.

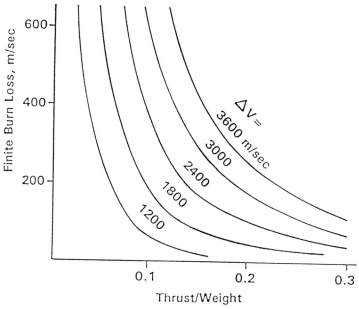

Fig. 3.2   Finite burn loss (for the Venus orbit insertion of a specific spacecraft).

Figure 3.2 is not generally applicable; each suspect situation requires a numerical integration if an accurate evaluation is to be made. Low thrust-to-weight ratios are to be avoided; situations in which thrust-to-weight is less than about 0.5 should be analyzed. Finite burn losses can be reduced by steering to hold the thrust vector on the spacecraft velocity vector.

3.2   Changes in Orbit

Orbital maneuvering is based on the fundamental principle that an orbit is uniquely determined by the position and velocity at any point. Conversely, changing the velocity vector at any point instantly transforms the trajectory to correspond to the new velocity vector, as shown in Fig. 3.3. Any conic can be converted to any other conic by adjusting velocity; a spacecraft travels on the trajectory defined by its velocity. Circular trajectories can be converted to ellipses; ellipses can be changed in eccentricity; circles or ellipses can be changed to hyperbolas—all by adjusting velocity. In Fig. 3.3,

$$\Delta V = V_p(\text{ellipse}) - V(\text{circle}) \qquad (3.4)$$

where

ΔV = velocity change required for the orbit change maneuver
$V(\text{circle})$ = spacecraft velocity at any point on a circular orbit
$V_p(\text{ellipse})$ = spacecraft velocity at the periapsis of the elliptical orbit, the point of closest approach to the central body

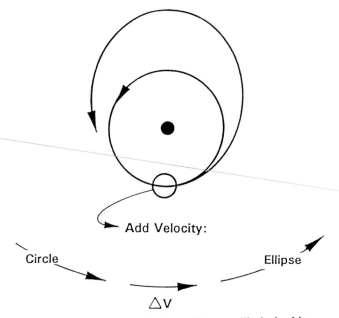

Fig. 3.3   Changing a circular orbit to an elliptical orbit.

The equations defining velocity and period of a circular orbit are:
Velocity, V:

$$V = \sqrt{\frac{\mu}{r}} \tag{3.5}$$

Period, P:

$$P = 2\pi \sqrt{\frac{r^3}{\mu}} \tag{3.6}$$

The relations governing an elliptical orbit are:
Velocity, V:

$$V = \sqrt{\frac{2\mu}{r} - \frac{\mu}{a}} \tag{3.7}$$

$$r_p V_p = r_a V_a \tag{3.8}$$

Eccentricity, e:

$$e = \frac{c}{a} \tag{3.9}$$

$$e = \frac{r_a}{a} - 1 \tag{3.10}$$

$$e = \frac{r_a - r_p}{r_a + r_p} \tag{3.11}$$

Period, P:

$$P = 2\pi\sqrt{\frac{a^3}{\mu}} \qquad (3.12)$$

Radius of apoapsis, r_a:

$$r_a = a(1 + e) \qquad (3.13)$$
$$r_a = 2a - r_p \qquad (3.14)$$

Radius of periapsis, r_p:

$$r_p = a(1 - e) \qquad (3.15)$$
$$r_p = 2a - r_a \qquad (3.16)$$

Semimajor axis, a:

$$a = \frac{r_a + r_p}{2} \qquad (3.17)$$
$$a = \frac{r_p}{1 - e} \qquad (3.18)$$

In Eqs. (3.7)–(3.18):

V = spacecraft velocity at the point on an orbit corresponding to the radius, r
r = radius at the point of interest, measured from the center of mass of the central body (usually Earth) to the spacecraft position
a = semimajor axis of the orbit, one-half of the long dimension of an ellipse
μ = gravitational parameter of the central body (see Appendix B for values)
e = orbital eccentricity, the ratio of the minor and major axes of an ellipse
r_p = periapsis radius, the minimum radius on an elliptical orbit
r_a = apoapsis radius, the maximum radius on an elliptical orbit

Example 3.1: Simple Coplanar Orbit Change

Consider an initially circular low Earth orbit at 300-km altitude. What velocity increase would be required to produce an elliptical orbit 300 × 3000 km in altitude? From Appendix B,

$$\mu = 398,600.4 \text{ km}^3/\text{s}^2$$

The velocity on the initially circular orbit is, from Eq. (3.5),

$$V = \sqrt{\frac{398,600.4}{(300 + 6378.14)}} = 7.726 \text{ km/s}$$

The semimajor axis on the final orbit is, from Eq. (3.17),

$$a = \frac{(300 + 6378) + (3000 + 6378)}{2} = 8028 \text{ km}$$

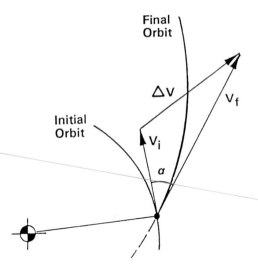

Fig. 3.4   Generalized coplanar maneuver.

The velocity at periapsis is, from Eq. (3.7),

$$V_p = \sqrt{\frac{2(398,600)}{6678} - \frac{398,600}{8028}} = 8.350 \text{ km/s}$$

For this maneuver, the propulsion system must provide a velocity increase $8.350 - 7.726 = 0.624$ km/s. The velocity would be increased at the point of desired periapsis placement. For a spacecraft weighing 1500 lb empty, the propellant required for this maneuver (if a delivered specific impulse of 310 s is assumed) is, from Eq. (3.3),

$$W_p = 1500\left\{\exp\left[\frac{(0.624)(3280.84)}{(32.174)(310)}\right] - 1\right\} = 341.8 \text{ lb}$$

Velocity changes, made at periapsis, change apoapsis radius but not periapsis radius, and vice versa; the radius at which the velocity is changed remains unchanged. As you would expect, the plane of the orbit in inertial space does not change as velocity along the orbit is changed. Orbital changes are a reversible process.

Figure 3.4 shows the general coplanar maneuver, which changes the initial orbit velocity V_i to an intersecting coplanar orbit with velocity, V_f. The velocity on the final orbit is equal to the vector sum of the initial velocity and the velocity change vector. Applying the cosine law yields

$$\Delta V = V_i^2 + V_f^2 - 2V_i V_f \cos\alpha \qquad (3.19)$$

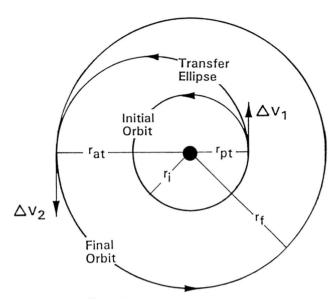

Fig. 3.5 The Hohmann transfer.

where

ΔV = Velocity added to a spacecraft initial velocity to change the
initial orbit to the final orbit; ΔV can be either positive or negative
V_i = Velocity on the initial orbit at the point of intersection of the two orbits
V_f = Velocity on the final orbit at the point of intersection of the two orbits
α = Angle between vectors V_i and V_f

The transfer can be made at any intersection of two orbits. Note that the least velocity change is necessary when the orbits are tangent and α is zero.

Hohmann Transfer

Suppose you need to transfer between two nonintersecting orbits. How can you do it? The Hohmann transfer, devised by Walter Hohmann[11] in 1925, employs an elliptical transfer orbit that is tangent to the initial and final orbits at the apsides. When it can be used, it is an energy-efficient orbital transfer. The Hohmann transfer is shown in Fig. 3.5. To design a Hohmann transfer, set the periapsis radius of the transfer ellipse equal to the radius of the initial orbit and the apoapsis radius equal to the radius of the final orbit.

$$r_{pt} = r_i \qquad (3.20)$$
$$r_{at} = r_f \qquad (3.21)$$

With these two radii set, the transfer ellipse is defined. Two velocity increments are required to accomplish the transfer: one to change the initial velocity of the spacecraft to the velocity on the transfer ellipse and a second to change from the velocity on the transfer ellipse to the velocity on the final orbit:

$$V_1 = V_{pt} - V_i \tag{3.22}$$
$$V_2 = V_{at} - V_f \tag{3.23}$$

where

V_{pt} = periapsis velocity on the transfer ellipse
V_{at} = apoapsis velocity on the transfer ellipse
V_i = spacecraft velocity on the initial orbit
V_f = spacecraft velocity on the final orbit

A transfer between two circular orbits is shown and described, but the transfer could as well have been between elliptical orbits. Similarly, the transfer could have been from the high to the low orbit.

Example 3.2: Hohmann Transfer

Design a Hohmann transfer from a circular Earth orbit of radius 8000 km to a circular orbit of radius 15,000 km.

The velocity on the initial orbit is, from Eq. (3.5),

$$V = \sqrt{\frac{398,600}{8000}} = 7.059 \text{ km/s}$$

Similarly, the velocity on the final orbit is 5.155 km/s.

The semimajor axis of the transfer ellipse is, from Eq. (3.17),

$$a = \frac{8000 + 15,000}{2} = 11,500 \text{ km}$$

The velocity at periapsis of the transfer ellipse is, from Eq. (3.7),

$$V_p = \sqrt{\frac{2(398,600)}{8000} - \frac{398,600}{11,500}} = 8.062 \text{ km/s}$$

Similarly, the velocity at apoapsis is

$$V_a = \sqrt{\frac{2(398,600)}{15,000} - \frac{398,600}{11,500}} = 4.300 \text{ km/s}$$

Another, and possibly easier, way to get velocity at apoapsis from Eq. (3.8) is

$$V_a = \frac{(8000)(8.062)}{15,000} = 4.300 \text{ km/s}$$

The velocity change required to enter the transfer orbit is

$$V_1 = 8.062 - 7.059 = 1.003 \text{ km/s}$$

Similarly, the velocity change to circularize is 0.855 km/s, and the total velocity change for the transfer is 1.858 km/s.

The efficiency of the Hohmann transfer comes from the fact that the two velocity changes are made at points of tangency between the trajectories.

3.3 Plane Changes

So far, we have considered only in-plane changes to orbits. What happens when a spacecraft must change orbit planes to accomplish its mission? An impulse is applied perpendicular to the orbit plane at the intersection of the initial and final orbit planes, as shown in Fig. 3.6. By inspection of Fig. 3.6 or, from the cosine law, it is clear that

$$\Delta V = 2V_i \sin \frac{\alpha}{2} \tag{3.24}$$

where

V = velocity change required to produce a plane change.
V_i = velocity of the spacecraft on the initial orbit at the point of
     intersection of the initial and final orbit planes
α = the angle of the plane change

Spacecraft speed is unaltered by the plane change, and orbit shape is unaffected; therefore, eccentricity, semimajor axis, and radii are unchanged.

Plane changes are expensive on a propellant basis. A 10-deg plane change in low Earth orbit would require a velocity change of about 1.4 km/s. For a 1000-lb spacecraft, this plane change would require 585 lb of propellant, if one assumes an I_{sp} of 310 s. Equation (3.24) shows that it is important to change planes through

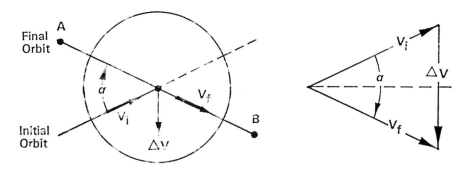

$$|\,V_i\,| = |\,V_f\,|$$

Fig. 3.6   Plane change maneuver.

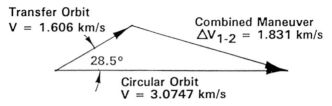

Fig. 3.7   Combined maneuver. For separate maneuvers, plane change maneuver
$\Delta V = 0.791$ km/s; circularization maneuver $\Delta V = 1.469$ ms; total $\Delta V = 2.260$ km/s.

the smallest possible angle and at the lowest possible velocity. The lowest possible
velocity occurs at the longest radius, that is, at apoapsis.

Combined Maneuvers

Significant energy savings can be made if in-plane and out-of-plane maneuvers
can be combined. The savings come from the fact that the hypotenuse of a vector
triangle is always smaller than the sum of the two sides. The maneuvers necessary
to establish a geosynchronous orbit provide an excellent example of combined
maneuver savings, as shown in Fig. 3.7.

The final two maneuvers are an in-plane orbit circularization and a 28.5-deg
plane change. The plane change maneuver is at a velocity of 1.606 km/s and would
require 0.791 km/s. The circularization maneuver would change the velocity at
apoapsis on the transfer ellipse to the circular orbit velocity of 3.0747 km/s; this
maneuver would require 1.469 km/s. The total requirement for velocity change
for the two maneuvers, conducted separately, would be 2.260 km/s. The two
maneuvers combined require a velocity change of only 1.831 km/s, a saving of
0.429 km/s. Watch for opportunities to combine maneuvers. Every plane change
is an opportunity of this kind.

3.4   Orbit Trim

Orbit trims are used to remove residual errors from a recently established
orbit. The errors could be caused by a launch vehicle or the spacecraft. Any
orbit parameter, radius, semimajor axis, eccentricity, or inclination may require
correction. For each error, it is necessary to perform a small orbit change, Hohmann
transfer, or plane change.

Launch vehicle injection accuracy is based on performance data and is reason-
ably well understood. Spacecraft missions vary in tolerance to errors. Geosynchronous and planetary spacecraft require close orbital tolerance. The Sputnik
mission would be possible in almost any orbit. Errors caused by the spacecraft
must be estimated by analysis of the particular spacecraft.

In addition to launch errors, orbital errors are introduced continuously by the
Earth's bulge, the attraction of the sun, the attraction of the moon, and solar
pressure. The spacecraft propulsion system must also make periodic corrections
for these perturbations. These correction maneuvers are called stationkeeping and
are performed as already described. The tighter the position tolerance, the more
stationkeeping maneuvers there are. The analysis of orbit trim requirements is
beyond the scope of this book (and of most preliminary studies). From its earliest

inception, however, a spacecraft program should carry an estimated propellant budget for orbit trims.

3.5 Repositioning

It may be necessary to change the position of the satellite. This is a common process for geosynchronous satellites and for spacecraft performing in a constellation. Repositioning is done by increasing (or decreasing) the velocity of the spacecraft to produce a more (or less) elliptical orbit. If the velocity is increased, the position of the impulse will become the perigee; if the velocity is decreased, the position of the decrease will become the apogee.

The velocity change (and hence propellant) required for repositioning can be accurately calculated by designing specific maneuvers and estimating the number required.

Example 3.3: Reposition

Consider a geosynchronous spacecraft that is required to reposition by 2-deg, counter to the velocity vector (westward), in a maneuvering time of one sidereal day (one orbit). Figure 3.8 shows the repositioning maneuver.

The elements of a geosynchronous orbit are

$$r = 42,164.17 \text{ km (circular)}$$
$$P = 86,164.09 \text{ s}$$
$$V = 3.07466 \text{ km/s}$$

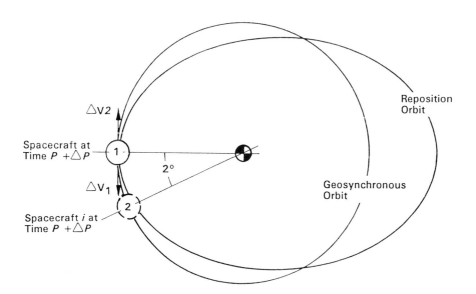

Fig. 3.8   Repositioning maneuver.

It is necessary to design a reposition ellipse tangent to the geosynchronous orbit with a longer period, such that the 2 deg reposition will occur in one orbit.

It is helpful to visualize a placeholder, spacecraft i, which occupies the position on the geosynchronous orbit while the real spacecraft goes into the reposition orbit. The placeholder returns to the point of injection (point 1 on Fig. 3.8) in one orbit period of P s. The spacecraft returns to point 1 in $P + \Delta P$ s. The reposition orbit is designed such that the placeholder will rotate exactly 2 deg beyond point 1 by the time the spacecraft arrives at point 1. ΔP is equal to the time required for 2 deg of motion on a geosynchronous orbit, which is

$$\Delta P = \frac{(2)(86{,}164.09)}{360} = 478.689 \text{ s}$$

The period for the spacecraft on the elliptical reposition orbit is

$$P = 86{,}164.09 + 478.689 = 86{,}642.78 \text{ s}$$

Rearranging Eq. (3.12) yields

$$a = \sqrt[3]{\frac{P^2\mu}{4\pi^2}} \tag{3.25}$$

and the semimajor axis of the reposition orbit is

$$a = \sqrt[3]{\frac{(86{,}642.78)^2(398{,}600)}{4\pi^2}} = 42{,}320 \text{ km}$$

The velocity at periapsis of an elliptical orbit with $a = 42{,}320.14$ km is, from Eq. (3.7),

$$V = \sqrt{\frac{2(398{,}600)}{42164} - \frac{398{,}600}{42320}} = 3.08032$$

Thus, the velocity change to place the spacecraft on the reposition ellipse is a reduction of $3.07466 - 3.08032 = 5.66$ m/s. In one lap, in the reposition ellipse, the spacecraft will be 2 deg of arc behind its old position. At that point exactly, the spacecraft velocity must be increased to its old value of 3.07466 km/s to replace it on the geosynchronous orbit. Thus, equal and opposite burns are required for a total of 11.32 m/s.

An excellent description of orbit errors, repositioning, and stationkeeping for geosynchronous spacecraft is given by Agrawal.[12]

3.6 Attitude Maneuvers

Some, if not all, of the thrusters on a spacecraft are devoted to attitude control. Since these motors must restart frequently, only the fluid motors (bipropellant, monopropellant, cold-gas) are candidates.

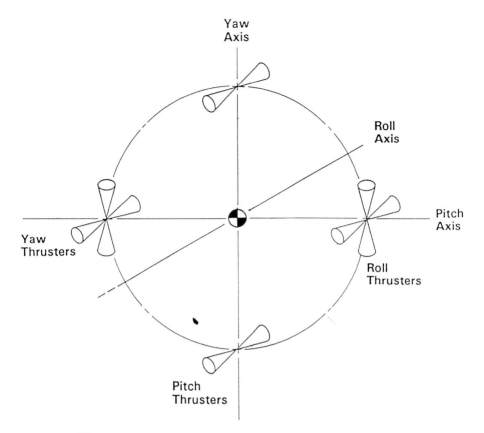

Fig. 3.9   Typical thruster installation for a three-axis spacecraft.

In a three-axis-stabilized system, an attitude maneuver consists of rotations about each of the spacecraft axes. Figure 3.9 shows a typical thruster installation for a three-axis-stabilized spacecraft. The thrusters are arranged so that the torques applied to the spacecraft are pure couples (no translational component).

Figure 3.10 shows thruster installation for a typical spinning spacecraft. In a spin-stabilized spacecraft, attitude maneuvers consist of translations for station-keeping and repositioning, reorienting the spin axis, and adjusting spin velocity.

Applying Torque to a Spacecraft

The elemental action in any rotational maneuver is applying a torque to the spacecraft about an axis. To apply torque, spacecraft thrusters are fired in pairs, producing a torque of

$$T = nFL \tag{3.26}$$

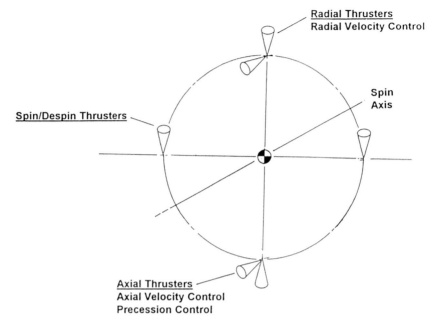

Fig. 3.10   Typical thruster installation for a spinning spacecraft.

where

T = torque on the spacecraft, ft-lb

F = thrust of a single motor, lb

n = number of motors firing, usually two; must be a multiple of two for pure rotation

L = radius from the vehicle center of mass to the thrust vector, ft

From kinetics,

$$\theta = \tfrac{1}{2}\alpha t_b^2 \tag{3.27}$$

$$\alpha = \frac{T}{I_v} \tag{3.28}$$

$$\omega = \alpha t_b \tag{3.29}$$

$$H = I_v\omega \tag{3.30}$$

$$H = T t_b \tag{3.31}$$

where

θ = angle of rotation of the spacecraft, rad

ω = angular velocity of the spacecraft, rad/s

α = angular acceleration of the spacecraft during a firing, rad/s^2

I_v = mass moment of inertia of the vehicle, slug-ft

t_b = duration of the burn

H = change of spacecraft angular momentum during the firing, slug-ft/s

During the burn, the angular acceleration of the spacecraft will be, from Eq. (3.28),

$$\alpha = \frac{nFL}{I_v} \qquad (3.32)$$

When the thrusters are shut down, the vehicle will have turned

$$\theta = \frac{nFLt_b^2}{2I_v} \qquad (3.33)$$

At shutdown, acceleration goes to zero and the spacecraft is left rotating at a velocity of ω. From Eqs. (3.28) and (3.29),

$$\omega = \frac{nFL}{I_v}t_b \qquad (3.34)$$

From Eq. (3.31), the angular momentum produced by a single firing is

$$H = Tt_b \qquad (3.35)$$

or

$$H = nFLt_b \qquad (3.36)$$

The propellant consumed during the burn is

$$W_p = \frac{nFt_b}{I_{sp}} \qquad (3.37)$$

or

$$W_p = \frac{H}{LI_{sp}} \qquad (3.38)$$

Equation (3.38) shows the advantage of a long moment arm. Expendable propellant can be saved by increasing moment arm. The maximum moment arm is constrained in a surprising way: by the inside diameter of the launch vehicle payload fairing. Table 3.2 lists the fairing inside diameter (i.d.) for some U.S. launch vehicles. It is not uncommon to take advantage of the full dynamic envelope (maximum available diameter allowing for dynamic flexing under load).

One-Axis Maneuver

A maneuver about one axis consists of three parts: 1) angular acceleration, 2) coasting, and 3) braking. Angular acceleration is produced by a thruster pair firing; braking is caused by a firing of the opposite pair. Figure 3.11 shows a one-axis maneuver.

The total angle of rotation is

$$\theta_m = \theta \text{ (accelerating)} + \theta \text{ (coasting)} + \theta \text{ (braking)}$$

Table 3.2   Launch vehicle shroud diameters
(from Ref. 13)

Launch vehicle	Fairing i.d., ft
Atlas	9.6 or 12
Delta	8.3 or 10
Space Shuttle	15
Titan II	10
Titan III	13.1
Titan IV	16.7

The rotation during coasting is

$$\theta = \omega t_c \qquad (3.39)$$

where t_c is the duration of the coasting in seconds. Using Eqs. (3.27) and (3.28), the coasting rotation angle is

$$\theta = \frac{nFL}{I_v} t_b t_c \qquad (3.40)$$

The rotation during acceleration or braking is given by Eq. (3.33); therefore, the total rotation during the maneuver is

$$\theta_m = \frac{nFL}{I_v} t_b^2 + \frac{nFL}{I_v} t_b t_c \qquad (3.41)$$

or

$$\theta_m = \frac{nFL}{I_v} \left(t_b^2 + t_b t_c \right) \qquad (3.42)$$

Note that t_b is the burn time for either of the two burns and that the maneuver time is

$$t_m = t_c + 2t_b \qquad (3.43)$$

The minimum time for a spacecraft to rotate through a given angle is a fully powered maneuver with zero coast time:

$$t_m = 2\sqrt{\frac{\theta_m I_v}{nFL}} \qquad (3.44)$$

The thrust level required for each thruster to perform a given maneuver in a given time is

$$F = 4\frac{\theta_m I_v}{nLt_m^2} \qquad (3.45)$$

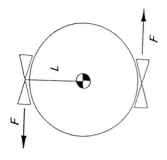

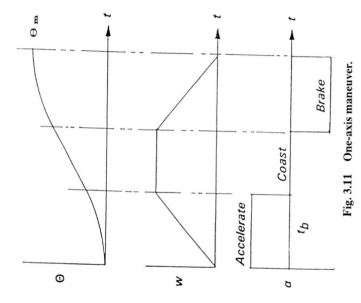

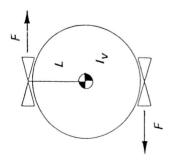

Fig. 3.11   One-axis maneuver.

The propellant required for a one-axis maneuver is twice the single burn consumption given by Eq. (3.37),

$$W_p = 2\frac{nFt_b}{I_{sp}} \qquad (3.46)$$

Virtually all maneuvers are three-axis maneuvers. Equation (3.46) must be applied to each axis to determine the total propellant required. Most maneuvers come in pairs, a maneuver to commanded attitude and a maneuver back to normal attitude. Therefore, the propellant required for a complete maneuver is usually twice that obtained from a three-axis application of Eq. (3.46).

Example 3.3: One-Axis Maneuver

Find the minimum time required for a spacecraft to perform a 90-deg turn about the z axis with two thrusters if the spacecraft has the following characteristics:

    Moment of inertia about the z axis = 1500 slug-ft
    Moment arm = 5 ft
    Thrust of each engine = 2 lb

and

$$\theta_m = \frac{\pi}{2} = 1.5708 \text{ rad}$$

From Eq. (3.44),

$$t_m = 2\sqrt{\frac{(1.5708)(1500)}{(2)(2)(5)}}$$
$$t_m = 21.708 \text{ s}$$

How much propellant was consumed by the maneuver if the $I_{sp} = 190$?

$$W_p = \frac{nFt_m}{I_{sp}} \qquad (3.47)$$

$$W_p = \frac{(2)(2)(21.708)}{190} = 0.4570 \text{ lb}$$

Note that this example can also be worked by the PRO software.

Precession of Spin Axis

Spin-stabilized spacecraft can maneuver by precession of the spin axis. Reorientation of the spin axis constitutes a change in the direction of the momentum vector, as shown in Fig. 3.12.

Putting a torque on the spacecraft rotates the momentum vector through angle ϕ:

$$\phi \approx \frac{H_a}{H_i} = \frac{nFLt}{I_y\omega} \qquad (3.48)$$

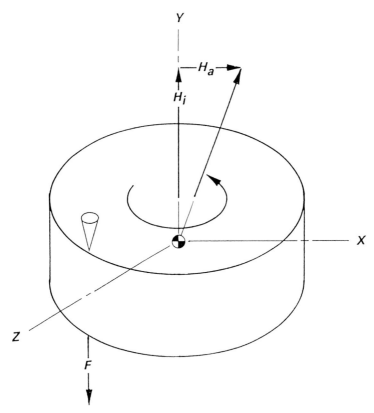

Fig. 3.12 Precession of spin axis.

where

H_a = momentum vector added, slug-ft/s
H_i = initial spin momentum, slug-ft/s
I_y = moment of inertia about the spin axis, slug-ft
ϕ = precession angle, rad
ω = angular velocity of the spacecraft, rad/s

Note that the pulse width t must be short compared to the period of spin. A pulse during an entire revolution would result in $H_a = 0$.

The spin axis will continue to precess until a second pulse of equal magnitude and opposite direction is fired. The spin axis can be repositioned by selecting the timing of the second pulse. The propellant consumed in a single pulse is given by Eq. (3.37).

Example 3.4: Precession of Spin Axis

What burn time, or pulse width, is required to precess a spacecraft spin axis by 3-deg (0.05236 rad) under the following conditions:

Thrust = 2 lb

Moment arm = 4 ft

Spacecraft spin rate = 2 rpm (0.2094 rad/s)

Moment of inertia = 1100 slug-ft^2

Specific impulse = 185 s

Two pulses are required to precess the spin axis; both pulses are parallel to the spin axis. The first pulse is used to cause nutation at an angle of one-half the desired precession. The second pulse stops the nutation and provides the remaining half of the desired angle. From Eq. (3.48), the burn time of either pulse is

$$t_b = \frac{\phi I_v \omega}{2n F L}$$

$$t_b = \frac{(0.05236)(1100)(0.2094)}{(2)(1)(2)(4)} = 0.7538 \text{ s}$$

The total propellant consumed by both burns is

$$W_p = 2\frac{n F t_b}{I_{sp}}$$

$$W_p = 2\frac{(1)(2)(0.7538)}{185} = 0.01630 \text{ lb}$$

Note that this example can also be worked by the PRO software.

3.7 Limit Cycles

Propulsion is frequently used to control spacecraft attitude in a limit-cycle mode. In this mode, attitude is allowed to drift until a limit is reached; then, thrusters are used for correction.

Without External Torque

A limit cycle without external torque swings the spacecraft back and forth between preset angular limits, as shown in Fig. 3.13 for one axis. When the spacecraft drifts across one of the angular limits θ_L, the attitude-control system fires a thruster pair for correction. The spacecraft rotation reverses and continues until the opposite angular limit is reached, at which time the opposite thruster pair is fired. It is important that the smallest possible impulse be used for the corrections because the impulse must be removed by the opposite thruster pair.

The total angle of rotation θ can be found by adapting Eq. (3.42). Note that the pulse width P_w, shown in Fig. 3.13, is twice the duration of t_b, as discussed in conjunction with one-axis maneuvers.

$$\theta = \frac{n F L}{I_v}\left(\frac{P_w^2}{4} + \frac{t_c P_w}{2}\right) \qquad (3.49)$$

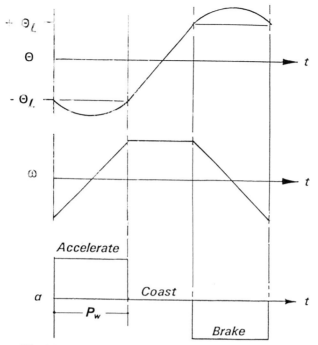

Fig. 3.13 Limit-cycle motion without external torque.

The limit settings $\pm\theta_L$ are one-half of the coasting angle, from Eq. (3.40),

$$\theta_L = \frac{nFL}{4I_v}P_w t_c \qquad (3.50)$$

Each cycle includes two pulses; the propellant consumed per cycle is

$$W_p = 2\frac{nFP_w}{I_{sp}} \qquad (3.51)$$

Equation (3.51) shows the importance of low thrust, short burn time, and high specific impulse in pulsing operation. Pulsing engines are characterized by minimum impulse bit I_{min}, where

$$I_{min} = F(P_w)_{min}$$

The minimum pulse width $(P_w)_{min}$ is a characteristic of a given thruster/valve combination. Table 3.3 shows typical pulsing performance of the propulsion system types. The pulsing I_{sp} shown in Table 3.3 is for cold engines (minimum duty cycle).

In a limit cycle, the propellant consumed per unit time is a key issue. The length of a cycle is

$$t_{cy} = 2P_w + 2t_c \qquad (3.52)$$

Table 3.3 Representative pulsing performance

	Min thrust, lb	Min impulse bit, lb-s	Pulsing I_{sp}, s
Cold-gas—helium	0.01	0.0001–0.0002	80
Cold-gas—nitrogen	0.01	0.0001–0.0002	50
Monopropellant—N_2H_4	0.1	0.001–0.002	120
Bipropellant—N_2O_4/MMH	2	0.015–0.030	120

From Eq. (3.50), the coast time is

$$t_c = \frac{4I_v\theta_L}{nFLP_w} \quad (3.53)$$

Thus, if minimum impulse bits are used, the length of a cycle is

$$t_{cy} = \frac{8I_v\theta_L}{nLI_{min}} + 2P_w \quad (3.54)$$

Note that the burn time is negligible (milliseconds) compared to the coast time (seconds) and can be neglected. This is also a conservative assumption. The propellant consumption per unit time is

$$\dot{W}_p = \frac{n^2 I_{min}^2 L}{4I_{sp}I_v\theta_L} \quad (3.55)$$

where $\dot{W}_p$ = propellant consumption per unit mission time in pounds per second.
Equation (3.55) shows the desirability of wide control band, low minimum impulse bit, and high I_{sp}.

Example 3.5: Limit-Cycle Operation

A spacecraft with 2000-slug-ft^2 inertia uses 1.5-lb thruster pairs mounted at a radius of 7 ft from the center of mass. For limit-cycle control to $\theta_L = 0.5$ deg (0.008727 rad), what is the propellant consumption rate if the I_{sp} is 170 s? The pulse duration is 30 ms, and there are no external torques.
The propellant consumed per cycle is

$$W_p = 2\frac{(2)(1.5)(0.03)}{170} = 0.0010588 \text{ lb/cycle}$$

The duration of a cycle is

$$t_{cy} = \frac{(8)(2000)(0.0087266)}{(2)(7)(1.5)(0.030)} + 2(0.030) = 221.69 \text{ s}$$

The propellant consumption rate is

$$\dot{W}_p = \frac{0.0010388}{221.69} = 4.8 \times 10^{-6} \text{ lb/s}$$

External Torque

Spacecraft are subject to a number of external torques:
1) Gravity gradient.
2) Solar pressure.
3) Magnetic field.
4) Aerodynamic.
Aerodynamic torques are proportional to atmospheric density and, hence, are important for low Earth orbits and rapidly become negligible in higher orbits. Solar pressure is the dominant disturbance above geosynchronous altitudes. Gravity gradient torques are dependent on spacecraft mass and mass distribution. Gravity gradients have been used to stabilize spacecraft. Propulsion requirements evolve from the necessity to correct for these torques. Wertz and Larson[14] provide excellent simplified methods for estimating external torques; these are summarized in Table 3.4. The equations in Table 3.4 are to be used to estimate relative magnitudes only. Note that solar radiation pressure T_{sp} is highly dependent on the type of surface being illuminated. A surface is either transparent, absorbent, or a reflector, but most surfaces are a combination of the three. Reflectors are classified as either specular (mirrorlike) or diffuse. In general, solar arrays are absorbers and the spacecraft body is a reflector.

One-Sided Limit Cycle

With an external torque on the spacecraft, rotation occurs until a limit line is reached and a thruster pair is fired for correction. If the limit lines are wider than a certain value, a one-sided limit cycle occurs. Figure 3.14 shows a one-sided limit cycle.

In Fig. 3.14, when the rotation reaches one of the limit lines, a thruster pair is pulsed. The unbalanced torque reverses the rotation, and the thrusters are fired to start another cycle. In effect, the unbalanced torque has replaced one of the thruster pairs.

For this cycle, the momentum supplied by the propulsion system exactly equals the momentum induced by the external torque:

$$H = T_x t \qquad (3.56)$$

where

T_x = external torque on the spacecraft, ft-lb
t = mission duration, s

The propellant weight required to compensate for the external torque is

$$W_p = \frac{T_x t}{L I_{sp}} \qquad (3.57)$$

A one-sided limit cycle is clearly an efficient way to deal with an external torque. For the cycle to occur, the rotational limits must be wide enough to prevent a

Table 3.4   Simplified equations for external torques (from Ref. 14, p. 315)

Disturbance	Type	Influenced primarily by	Formula
Gravity gradient	Constant or cyclic, depending on vehicle orientation	Spacecraft geometry Orbit altitude	$T_g = \dfrac{3\mu}{r^3}\lvert I_z - I_y\rvert\theta$ where T_g is the max gravity torque; μ is the Earth's gravity constant (398,600 km^3/s^2), r the orbit radius, θ the max deviation of the z axis from vertical in radians; I_z and I_y are moments of inertia about z and y (or x, if smaller) axes.
Solar radiation	Constant force but cyclic on Earth-oriented vehicles	Spacecraft geometry Spacecraft surface area	The worst-case solar radiation torque $T_{sp} = P_s A_s L_s (1+q)\cos i$ is due to a specularly reflective surface, where P_s is the solar constant, 4.617×10^{-6} N/m^2; A_s is the area of the surface, L_s the center of pressure to center of mass offset, i the angle of incidence of the sun, and q the reflectance factor that ranges from 0 to 1; $q = 0.6$ is a good estimate.
Magnetic field	Cyclic	Orbit altitude Residual spacecraft magnetic dipole Orbit inclination	$T_m = 10^{-7} D B$ where T_m is the magnetic torque on the spacecraft, D the residual dipole of the vehicle in pole/centimeters, and B the Earth's magnetic field in gauss. B can be approximated as $2M/r^3$ for a polar orbit to half that at the equator. M is the magnetic moment, 8×10^{25} emu at Earth, and r is radius from dipole (Earth) center to spacecraft in centimeters.
Aerodynamic	Constant for Earth-oriented vehicle in circular orbit	Orbit altitude Spacecraft configuration	$T_a = \sum F_i L_i$ T_a is the summation of the forces F_i on each of the exposed surface areas times the moment arm L_i to the center of each surface to the center of mass, where $F = 0.5[\rho C_d A V^2]$ with F the force, C_d the drag coefficient (usually between 2.0 and 2.5), ρ the atmospheric density, A the surface area, and V the spacecraft velocity.

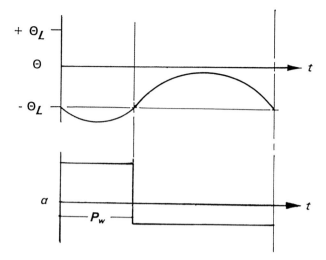

Fig. 3.14   One-sided limit cycle.

thruster firing from assisting the external torque. The rotation limit must be wider than

$$\theta_L > \frac{n^2 F^2 L^2 P_w^2}{16 I_v T_x} \qquad (3.58)$$

Forced Limit Cycle

A forced limit cycle, shown in Fig. 3.15, occurs when thrusters are fired in the direction of the external torque; that is, when the condition of Eq. (3.58) is not met.

The propellant consumed in a forced limit cycle is

$$W_p = \frac{I_v R^2 t}{L \theta_L I_{sp}} \qquad (3.59)$$

where

$R = 1/t_c$ = limit-cycle rate of the system, Hz
t = mission duration, s

It is important to avoid a forced limit cycle because of the substantial propellant consumption.

3.8   Reaction Wheel Unloading

A reaction wheel is a flywheel driven by a reversible dc motor. To perform a rotational maneuver with a reaction wheel, the flywheel is accelerated. The

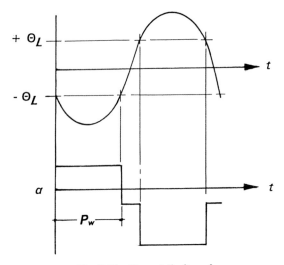

Fig. 3.15　Forced limit cycle.

spacecraft accelerates in the opposite direction. To do a maneuver with reaction wheels,

$$\theta = \frac{T t_m^2}{4 I_v} \qquad (3.60)$$

and torque is

$$T = \alpha_w I_w \qquad (3.61)$$

where subscript w refers to the properties of the reaction wheel. The angle of spacecraft rotation is

$$\theta = \frac{\alpha_w I_w t_m^2}{4 I_v} \qquad (3.62)$$

The increase in wheel speed is

$$\Delta \omega_w = \frac{\alpha_w t_m}{2} \qquad (3.63)$$

where ω_w is the wheel speed, rad/s.

When the spacecraft is returned to its initial position, the net change in wheel speed is zero for a frictionless wheel, which is the major advantage of reaction wheels. Eventually, however, unbalanced torques will load the wheel (bring it to maximum speed). The wheel can be unloaded by reversing the motor and holding the spacecraft in position with reaction jets. The momentum of the wheel, H, is

$$H = Tt = nFLt \qquad (3.64)$$

and

$$H = I_w \omega_w \tag{3.65}$$

The total impulse required to unload the reaction wheel is

$$I = nFt = \frac{I_w \omega_w}{L} \tag{3.66}$$

During unloading, the momentum will be transferred from the propulsion system. Assuming that wheel speed is brought to zero, the propellant required to unload is

$$W_p = \frac{I_w \omega_w}{L I_{\text{sp}}} \tag{3.67}$$

The time required to unload is

$$t = \frac{I_w \omega_w}{nFL} \tag{3.68}$$

When reaction wheels are controlling external torques, the momentum required for wheel unloading is equal to the momentum due to the external torque; Eq. (3.57) applies.

Example 3.6: Reaction Wheel Unloading

How much propellant does it take to unload one of the Magellan wheels, and how long does it take? The Magellan wheel characteristics are:

Maximum momentum = 27 N-m-s = 19.939 ft-lb-s
Maximum wheel speed = 4000 rpm = 418.879 rad/s

The thruster pair to be used has the following characteristics: thrust = 0.2 lb; moment arm = 7 ft; pulsing specific impulse = 150 s.

If it is assumed that the wheel is fully loaded, the propellant required to unload it is

$$W_p = \frac{19.9139}{(7)(150)} = 0.01896 \text{ lb}$$

The engine burn time required to unload is

$$t = \frac{19.9139}{(2)(0.2)(7)} = 7.112 \text{ s}$$

Note that this example can also be worked with the PRO software.

Problems

3.1   A 3000-lb spacecraft is in 280-km-altitude circular parking orbit. How much propellant is required to place it on a Hohmann transfer ellipse from the parking orbit to geosynchronous orbit ($r_a = 42,164$ km) if the propulsion system can deliver an I_{sp} of 310 s?

3.2   A 2800-lb Earth resources spacecraft has been placed in a 400 × 2000-km-altitude elliptical orbit. How much propellant is required to circularize the orbit at 400-km altitude if the propulsion system can deliver an I_{sp} of 290 s?

3.3   A spacecraft has a monopropellant propulsion system that delivers an I_{sp} of 225 s. How much propellant would be consumed to trim the orbit if a ΔV of 200 m/s were required and the spacecraft weighed 2000 lb at the end of the burn?

3.4   During its next flight, the Space Shuttle Columbia is required to make a 3-deg plane change in its 275-km-altitude circular orbit. The Columbia weighs 220,000 lb before the maneuver, and the propulsion system delivers an I_{sp} of 320 s. How much propellant must be loaded for the maneuver?

3.5   A spacecraft has the following characteristics:
  a) Thruster pair, each located at 6.5-ft radius from the center of mass.
  b) Moment of inertia of 4700 ft-lb/s^2 about the z axis.
  c) Thrust of each engine is 0.2 lb.
What is the minimum time for the spacecraft to maneuver through a 65-deg rotation about the z axis? How much propellant is consumed at a specific impulse of 185 s?

3.6   A spacecraft must be maneuvered through an angle of 60-deg in 30 s. The spacecraft inertia is 8000 slug-ft^2; the thrusters are located at a radius of 7 ft from the center of mass. What is the minimum thrust level for the thrusters?

3.7   A spacecraft with the following characteristics is in a limit-cycle operation about the z axis:
  Angular control requirement $= \pm 0.5$ deg
  Moment of inertia about $z = 12,000$ slug-ft^2
  Minimum impulse bit $= 0.01$ lb-s (0.5-lb engines with $t_{on} = 0.02$ s)
  Pulsing specific impulse $= 120$ s
  Two thrusters mounted 6 ft from the center of mass
If external torques are negligible, what is the time rate of fuel consumption?

3.8   You are sizing the thrusters for a new three axis–stabilized spacecraft. The driving requirement is the need to slew 90 deg about any axis in 5 min using 30-s burns. Twelve identical thrusters will be located on struts that place them 2 m from the center of mass. The nonredundant thrusters provide pure couples. What is the minimum thrust level you need? The expected moments of inertia are: $I_x = 1000$ kg/m^2, $I_y = 2000$ kg/m^2, and $I_z = 12,000$ kg/m^2.

3.9   An upper stage is designed to have a usable propellant load of 5000 lb and a burnout weight of 850 lb. If the stage is used to deliver a ΔV of 15,000 ft/s to a 600-lb payload, what is the vacuum specific impulse? Assume drag $= 0$ and vacuum conditions.

3.10   How much propellant is required for 24 h of roll-axis limit-cycle operation under the following conditions?
   Roll-axis moment of inertia $= 6500$ ft-lb/s^2
   Minimum impulse bit $= 0.001$ lb/s
   Radius, center of mass to thrusters $= 6$ ft
   Pointing accuracy required $= 0.01$ deg
   No external torque
   Specific impulse (pulsing) $= 130$ s

3.11   A 2120-lb communications satellite has attained an orbit in the equatorial plane; however, the apogee altitude is 41,756 km and the eccentricity is 0.0661. What is the minimum amount of monopropellant to place the satellite in a geosynchronous orbit? The final orbit altitude must be $35,786 \pm 10$ km, and the system specific impulse is 225 s.

3.12   What is the maximum velocity increase that an upper stage, operating in a vacuum, will impart to a spacecraft that weighs 7920 lb? The upper stage has the following characteristics:
   Stage dry weight $= 1326$ lb
   Usable propellant $= 25,320$ lb
   Unusable propellant $= 36$ lb
   Specific impulse $= 456$ s

4
Monopropellant Systems

A monopropellant system generates hot, high-velocity gas by decomposing a single chemical, a monopropellant. The concept is shown in Fig. 4.1. The monopropellant is injected into a catalyst bed, where it decomposes; the resulting hot gases are expelled through a converging-diverging nozzle generating thrust. A monopropellant must be a slightly unstable chemical that decomposes exothermically to produce a hot gas, as a number of chemicals do. Table 4.1 lists some of these.

A number of practical considerations, notably stability, thin the list in Table 4.1 Only three monopropellants have ever been used on flight vehicles: hydrazine, hydrogen peroxide, and propyl nitrate. Shock sensitivity eliminated propyl nitrate after limited use for jet engine starters. Hydrogen peroxide saw considerable service as a monopropellant in the 1940s to 1960s (starting with the V-2). The persistent problem with hydrogen peroxide is slow decomposition during storage. The decomposition products cause a continuous increase in pressure in the storage vessel and in water dilution of the propellant. The pressure rise complicates flight and ground tank design; the water dilution reduces performance.

Table 4.1 shows that hydrazine has desirable properties across the board, including the highest specific impulse of the stable chemical group. A serious limitation on the early use of hydrazine was ignition; hydrazine is relatively difficult to ignite. In the early years, hydrazine was ignited by injecting a start slug of nitrogen tetroxide. Once lit, the combustion was continued by a catalyst. All of the Rangers, Mariner 2 as well as Mariner 4, used a hydrazine start slug system. Using start slugs limited the number of burns to the number of slugs, two in the case of Mariner 4. Start slugs also complicate the system. The great advantage of monopropellants is the elimination of the oxidizer system. With the start slug, the oxidizer system is still necessary.

The need for a spontaneous catalyst led the Shell Development Company and the Jet Propulsion Laboratory to develop the Shell 405 iridium pellet catalyst bed in 1962. Almost unlimited spontaneous restart capability resulted. The spontaneous restart capability, along with relative stability, high performance, clean exhaust, and low flame temperature, has made hydrazine the only monopropellant in use today.

4.1 Monopropellant Hydrazine Thrusters

A monopropellant hydrazine thruster is shown in Fig. 4.2. The propellant flow into the chamber is controlled by a propellant valve that is an integral part of the thruster. The propellant is injected into a catalyst bed, where it decomposes into hydrogen, nitrogen, and ammonia. The gases are expelled through a converging-diverging nozzle producing thrust. The fundamental rocket engine

Table 4.1   Characteristics of some monopropellants (Reprinted from
H. Koelle, *Handbook of Astronautical Engineering*, McGraw-Hill, 1961.)

Chemical	Density	Flame temp, °F	C^*, fps	I_{sp}, s	Sensitivity
Nitromethane	1.13	4002	5026	244	Yes
Nitroglycerine	1.60	5496	4942	244	Yes
Ethyl nitrate	1.10	3039	4659	224	Yes
Hydrazine	1.01	2050	3952	230	No
Tetronitromethane	1.65	3446	3702	180	Yes
Hydrogen peroxide	1.45	1839	3418	165	No
Ethylene oxide	0.87	1760	3980	189	No
n-Propyl nitrate	1.06	2587	4265	201	Yes

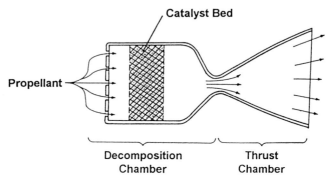

Fig. 4.1   Monopropellant concept.

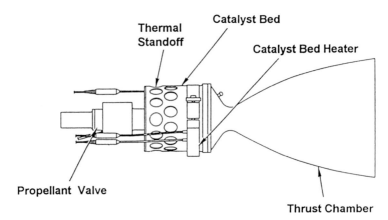

Fig. 4.2   Typical monopropellant thruster. (Courtesy Hamilton Standard.)

equation applies; the thrust generated is

$$F = W_p/g_c + (P_e - P_a)A_e \qquad (4.1)$$

The gas temperature is in the 2200°F range. High-temperature alloys can be used for the converging-diverging nozzle, and supplemental cooling is not required. The valve performance is an integral part of the thruster performance during pulsing, and the two are considered as a unit. Monopropellant thrusters have flown in sizes ranging from 0.1 to 600 lb and blowdown ratios up to 6. Pulse widths as low as 7 ms have been demonstrated[10] as have thrust levels in the 1000-lb class.

Steady-State Performance

The decomposition of hydrazine leads first to hydrogen and ammonia. The reaction is exothermic, and the adiabatic flame temperature is about 2600°F; however, the ammonia further decomposes into hydrogen and nitrogen. This reaction is endothermic and leads to a reduction in flame temperature and I_{sp}. It is desirable to limit the dissociation of ammonia as much as possible. Ammonia dissociation can be held to about 55% in current engine designs. Figure 4.3 shows the performance of anhydrous hydrazine as a function of percent ammonia dissociation. A steady-state, theoretical, vacuum specific impulse of about 240 s can be expected at an area ratio of 50:1. Real system specific impulse will be about 93% of theoretical. Specific impulse at other area ratios can be estimated from the ratio of thrust coefficients (see Chapter 2). The ratio of specific heats for the exhaust is about 1.27.

Pulsing Performance

Attitude-control applications of monopropellant hydrazine engines require operation over wide ranges of duty cycles and pulse widths. Such use makes the pulsing specific impulse and minimum impulse bit very important. Pulsing involves the performance of the propellant valve and feed tubing as well as the chamber itself. A typical pulse is shown in Fig. 4.4.

Pressure response time is the time (measured from the propellant valve actuation signal) required to reach an arbitrary percentage of steady-state chamber pressure. Response time is affected by:

1) The valve response characteristics.
2) The feed line hydraulic delay (time for the propellant to flow from the valve to the injector).
3) The ignition delay (time to wet the catalyst bed and for initial decomposition heat release to bring the catalyst to a temperature at which ignition becomes rapid).
4) The pressure rise time (time to decompose enough hydrazine to fill the void space in the reactor and heat the whole catalyst bed).

Valve response characteristics and feed line hydraulic delay vary for each specific design. Valves are available with response times better than 10 ms. The distance between the valve and injector must be minimized; however, the control of heat soak back to the valve sets a minimum length for the injector tube.

With proper injector design, ignition delay will be approximately 10 to 20 ms for catalyst and propellant temperatures in the range of 40° to 70°F. With a catalyst bed temperature in the range of 500°F, ignition delay will be approximately 1 to 2 ms.

Pressure rise time is a much larger fraction of the response time than ignition delay for most thrust chambers. Response times from valve signal to 90% of steady-state chamber pressure of 15 ms have been demonstrated with tail-off time (signal to 10%) of 20 ms.[15]

In Fig. 4.4, the response from signal to 90% thrust is 15 ms; the tail offtime is 20 ms. If the duty cycle is long, the engine will cool between pulses, and thrust

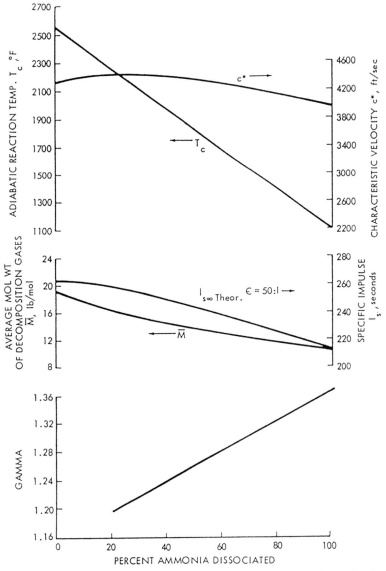

Fig. 4.3   Hydrazine performance vs ammonia dissociation (from Ref. 15, p. 23).

will not reach rated value because of energy losses to engine heating. If pulses are frequent and the engines hot, 100% thrust will be the rated thrust.

In the absence of test data, the minimum impulse bit can be estimated by

$$I_{min} \approx F t \qquad (4.2)$$

where

F = steady-state thrust reached, lb
t = pulse width, time from valve *on* to valve *off* s

The results of this approximation are compared to test data in Table 4.2. Note that, for infrequent pulses, the thruster will be cold, and full-rated thrust will not be reached. For such cases, the thrust level corresponding to the expected gas temperature should be used.

Figure 4.5 shows pulsing vacuum specific impulse for monopropellant hydrazine systems as a function of pulse width and engine offtime. Pulsing specific impulse is low at low duty cycles because energy is lost reheating the motor. Short pulses deliver low specific impulse for the same reason. Low duty cycles and short pulses in combination deliver specific impulse as low as 115 s, as shown in Fig. 4.5 Specific impulse is limited at low duty cycles by the performance of an ambient temperature bed. The performance of a thruster operating with a cold bed can be estimated by assuming the exhaust gas exits at the bed temperature.

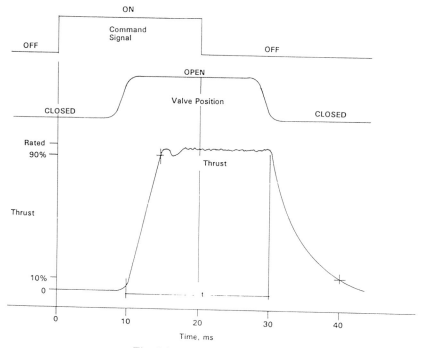

Fig. 4.4   A typical 20-ms pulse.

Table 4.2   Calculated and measured impulse bit

Rated thrust, lb	Pulse width, ms	Measured impulse	Ft, lb-s
130	22	2.4	2.9
130	66	7.5	8.6
40	22	1.1	0.9
40	66	3.0	2.6
7	20	0.11	0.14

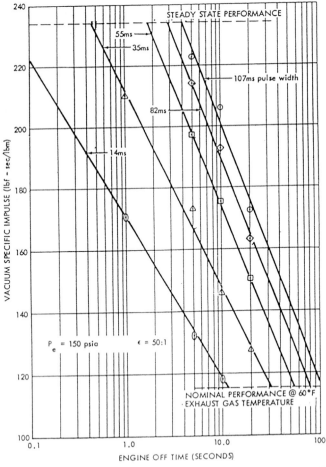

Fig. 4.5   Pulsing specific impulse of hydrazine (from Ref. 10, P. 65).

All of the hydrazine that reaches the catalyst bed is decomposed; there are no liquid losses on start or shutdown.

Negative Pulses

In a steady-state burn, it is often desirable to pulse off to create attitude-control torques. The Magellan and Viking spacecraft have used the pulse-off technique. In the Magellan case, the 100-lb thrusters were fired in steady state during the solid motor burn for Venus orbit insertion. (The 100-lb engines are aligned with the motor thrust vector.) Pulsing off was used for thrust vector control during the burn. A similar strategy was used by Viking Lander during the deorbit phase.

Negative impulse is the impulse lost during a pulse-off as compared to the impulse that would have been generated in steady state. Negative pulses are shaped differently from positive ones; however, the product of steady-state thrust and valve offtime is a satisfactory approximation of the negative impulse. Table 4.3 compares the multipulse average impulse of three engines in pulse-on and pulse-off modes.

Performance Accuracy and Repeatability

The steady-state performance variation is independent of inlet pressure and can be expected to be about:

Steady-state thrust, engine to engine, $\pm 5\%$

Steady-state I_{sp}, engine to engine, $\pm 2\%$

Pulsing performance is difficult to generalize. The early pulses in a pulse train will not be repeatable pulse to pulse. After 5 or 10 pulses, the impulse bit variation for a given engine will be about $\pm 3\%$ pulse to pulse for pulses longer than 1 s. As pulse width is shortened, the variability increases, as shown in Fig. 4.6. For

Table 4.3   Comparison of pulse-on and pulse-off modes

Rated thrust, lb	Mode	Pulse width, ms	Impulse bit, lb/s
130	Pulse-on	22	2.4
130	Pulse-off	22	2.8
130	Pulse-on	66	7.5
130	Pulse-off	66	10.7
40	Pulse-on	22	1.1
40	Pulse-off	22	0.6
40	Pulse-on	66	3.0
40	Pulse-off	66	2.1
7	Pulse-on	20	0.11
7	Pulse-off	20	0.11
7	Pulse-off	40	0.25
7	Pulse-off	100	0.75
7	Pulse-off	400	1.30

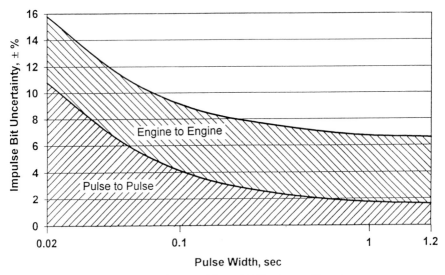

Fig. 4.6   Typical impulse bit variability for small engines.

a train of pulses, the impulse variability will be about ±10% pulse to pulse. The variability engine to engine is about ±5% and is additive. If you fired a number of 20-ms pulses on a number of different engines, the variation over the population of pulses would be about ±15%.

Thruster Life

The degradation of a monopropellant thruster occurs in the catalyst bed as a result of two mechanisms: 1) mechanical failure of the catalyst pellets and 2) reduction of catalytic activity caused by impurities on the surface of the pellets (poisoning). Poisoning is proportional to hydrazine flow rate; therefore, high-thrust motors have a shorter cycle life. Poisoning is reduced by the use of superpure or Viking-grade hydrazine. Voyager ground-test results confirmed the desirability of superpure hydrazine[16]; Magellan, Viking, and Voyager are among the spacecraft programs to use it.

Cold starts are responsible for most of the physical damage to catalyst pellets. The ignition delay associated with a cold start allows propellant to accumulate in the bed. When ignition occurs, an overpressure spike stresses the pellets. In addition, cold starts thermally stress the pellets. Maintaining the catalyst bed at an elevated temperature eliminates cold starts and substantially reduces the physical degradation of the pellets. The 0.1-lb thrusters on FLEETSATCOM are maintained at 315°C and are expected to achieve 1 million cycles.[12] The catalyst beds on Voyager were maintained at 200°C and qualified for 400,000 cycles. Approximately 400,000 pulses were accomplished during the flight.[16] Pulsing performance, as well as catalyst bed life, is improved by catalyst bed heaters.

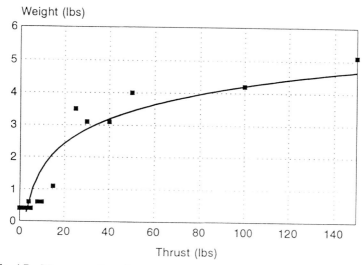

Fig. 4.7   Monopropellant thruster weight (data in part from Refs. 7 and 15).

Thruster Weight

A least-squares curve fit of the weight of nine different thruster/valve designs with thrust levels from 1 to 150 lb produces the following relation:

$$W_t = 0.34567F^{0.55235} \qquad (4.3)$$

Figure 4.7 shows the correlation; the correlation coefficient is 0.97.

For low thrust levels, the thruster weight approaches the valve weight, an effect that Eq. (4.5) will not predict. Use 0.4 lb as a minimum thruster/valve weight for low thrust levels. Note that Fig. 4.7 is for a thruster with single valves.

4.2   Electrothermal and Arcjet Thrusters

If the combustion products of hydrazine decomposition are increased in temperature, significant increases in specific impulse can be achieved. In this section, two methods of using electric power to increase temperature and specific impulse will be discussed: 1) electrothermal thrusters and 2) arcjet thrusters.

These thruster designs offer substantial increase in specific impulse and should be considered in the following situations:

1) The time line will allow time to precondition the thruster before firing.
2) Steady-state burns are required, as opposed to pulsing.
3) Relatively low thrust levels are required.
4) Substantial spacecraft power is available for short periods.

Electrothermal Thrusters

Electrothermal thrusters offer a viable alternative to the standard monopropellant thruster design. The design is different in two ways: Electrical energy is used

Table 4.4   Characteristics of Intelsat V electrothermal
thruster (data from Ref. 17)

Thrust	0.1 lb
Average I_{sp}	295 s
Total impulse	18,000 lb/s
Firing time	210 h
Total weight	0.78 lb
Area ratio	200
Overall dimensions	5.3 long, 2.6 wide,
	3.4 in diameter in.,
Inlet pressure	270 to 120 psia
Power	414 W

to 1) cause decomposition and 2) to increase gas temperature. The result is a significant increase in specific impulse. Figure 4.8 shows the electrothermal thrusters used on Intelsat V.

The performance and design characteristics of the motor are given in Table 4.4.

In Fig. 4.8, hydrazine is fed from the propellant valve into the decomposition chamber. There it is decomposed over a platinum thermal capacitance screen pack. The screen pack is heated to 1300°F prior to firing by electrical heaters strapped to the decomposition chamber. The hydrogen, nitrogen, and ammonia mixture leaves the decomposition chamber at a temperature of 1600–1800°F. Directly downstream of the decomposition chamber is a high-powered vortex heat exchanger. Exhaust gases are injected tangentially into the chamber to create a

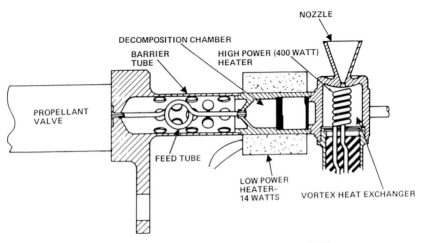

Fig. 4.8   Electrothermal thruster (from Ref. 17).

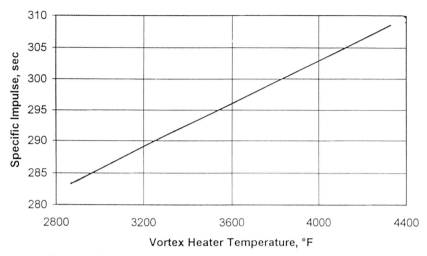

Fig. 4.9   Delivered I_{sp} vs vortex heater temperature (from Ref. 17).

vortex around the heat exchanger, where the gas temperature is raised to 3000–3500°F.[33] The specific impulse achieved by these thrusters is shown in Fig. 4.9 as a function of vortex heater temperature for vacuum conditions at an area ratio of 200:1.[17]

During flight, the decomposition heater is turned on about 15 min prior to firing. The vortex heater is turned on 15 to 30 s after the firing command, and turned off 5 to 10 s prior to the shutdown command. After shutdown, the decomposition heaters are turned off.

Arcjet Thrusters

Another form of electrothermal thrusters, arcjets, increase the temperature of the combustion gases by use of an electrical arc rather than an vortex heater.

The thruster, shown in Fig. 4.10, consists of a power control unit (PCU), an electrode assembly, and a gas generator and associated valves. Hydrazine enters the gas generator through thrust control valves; the gas generator decomposes the hydrazine using an iridium catalyst in much the same way as a conventional thruster. A gas manifold feeds the hot decomposition products into the electrode region; the downstream end of the manifold is shaped to impart a rotary motion to the gas. The anode forms the throat of the thruster through which the arc passes and where most of the heating occurs. The expansion area, also part of the anode, increases the gas velocity and provides for arc attachment grounding.

Arcjet thrusters are being qualified for N/S stationkeeping on AT&T's Telstar spacecraft; the performance of the Telstar thrusters is summarized in Table 4.5. The Telstar arcjet system mass is summarized in Table 4.6. With the arc turned off, the thrusters will perform as conventional engines with conventional I_{sp}. Arcjet starts are a multistep process:

1) The thrusters are fired in conventional mode for about 20 s to assure the presence of gas in the arc region.

Table 4.5   Telstar arcjet test results (from Ref. 18)

	Steady state (50 firings)	Duty cycle (1 h on, 0.5 h off)
Power level, W	1620	1620
Thrust, lb	0.05	0.05
Specific impulse, s	>520	>520
Life, h	1258	870
Starts	183	>900
Total impulse, N/S	978,000	
Blowdown Pressurization, psia	300–220	300–220

Table 4.6   Telstar arcjet mass

Component	Mass, kg
Thruster	1.0
Power conditioning unit	4.2
Cable/Connectors	0.8
Total	6.0

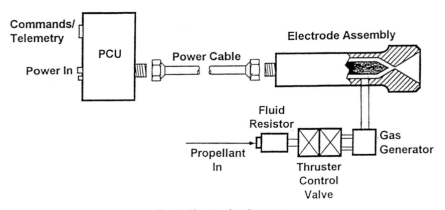

Fig. 4.10   Arcjet thruster.

2) The arc is started.
3) The arc transitions through the constrictor
4) Steady-state temperature and performance are established.

4.3 Propellant Systems

The purpose of the propellant system is to contain the propellants and to serve them to the engine on demand at the proper pressure, quality, and cleanliness. Hydrazine is the only monopropellant in use today. Anhydrous hydrazine is a clear, colorless hygroscopic liquid with a distinct, ammonialike odor. It is a stable chemical that can be stored for long periods without loss of purity. It is relatively insensitive to shock compared to other monopropellants.

Hydrazine is a strong reducing agent and it is toxic. Special preparation, equipment, and procedures are required for handling it. The properties of hydrazine and other propellants are given in Appendix B.

Propellant Inventory

The propellant inventory is a subdivided tabulation of the loaded propellant weight. Table 4.7 shows an example. Two activities dictate the required usable propellant: 1) the propellant required for mission maneuvers as dictated by the mission design and 2) the propellant required to control attitude. On Magellan, we kept track of the propellant for each of the major maneuvers plus attitude control because each of these change. Propellant for mission maneuvers is a product of the mission analysis calculations. For example, it takes a velocity change of about 2.25 km/s to move a spacecraft from a parking orbit to a geosynchronous orbit; this propellant is mission maneuver propellant. Attitude-control propellant is required for duty cycle operation, wheel unloading, or spacecraft pointing. (Chapter 3 shows how these propellant requirements are calculated.)

Some design groups calculate the propellant requirements under nominal conditions. If this is done, it is necessary to calculate the additional propellant necessary to complete the mission under worst-case conditions (item 3 in Table 4.7). It is preferable to calculate the propellant requirements using worst-case conditions, in which case item 3 is unnecessary.

Table 4.7 Propellant inventory

1) Propellant for mission maneuvers
2) Propellant for attitude control
3) Off-nominal allowance (optional)
4) Reserves
5) Subtotal: Usable propellant   XXXXXXX
6) Trapped propellant, 3% of usable
7) Outage (bipropellants only), 1% of usable
8) Loading uncertainty, 0.5% of usable
9) Total loaded propellant   XXXXXXX

Propellant reserves are a very valuable commodity. The more of the project reserves placed in propellant, the better. There are three primary reasons for this: 1) Propellant is usually the life-limiting expendable. 2) It is often desirable to make unplanned maneuvers in emergency conditions. 3) There are usually extended mission objectives that can be achieved after the primary mission—if there is propellant. Voyager, for example, visited Uranus and Neptune after the primary mission was over. Propellant reserve can be used in flight in many useful ways unlike weight margin, power margin, and most other margin types. The propellant reserve should be as much as you can stand and never less than 25% of usable in the early phases.

The sum of the mission propellant, attitude-control propellant, and reserves (and item 3, if necessary) is the usable propellant. The usable propellant is that portion of the propellant loaded that is actually burned; it is the loaded amount less the unusable. Not all of the propellant loaded can be used; a certain amount is trapped, outage, drop-out, or loading uncertainty.

Trapped propellant is the propellant remaining in the feed lines, tanks, and valves; hold up in expulsion devices; and retained as vapor in the system with the pressurizing gas. Trapped propellant is about 3% of the usable propellant.

Outage (bipropellants only) is caused by the difference between the mixture ratio loaded and the mixture ratio burned. There is always some of one propellant left when the other propellant is depleted. Outage depends on loading accuracy and on the burn-to-burn repeatability of engine mixture ratio. For initial estimates, 1% outage can be used; it takes statistical engine data and loading data, however, to make an informed estimate.

Drop-out (launch vehicle systems only) is the propellant left in the feed line when the propellant surface in the tank drops into the line.

Uncertainty in the measurement of propellant loading is about 0.5%. The uncertainty is added to the load to insure that the usable propellant can be no less than the worst-case requirement.

Zero-g Propellant Control

The location of the ullage bubble, or bubbles, becomes indeterminate in zero-*g* conditions unless the liquid and gas are positively separated. Figure 4.11 shows zero-*g* bubble shapes.

Unless the gas location is controlled, there is no way to guarantee gas-free liquid to the engine. Although an engine can tolerate small amounts of gas ingestion without failure, it is normal design practice to prevent any gas ingestion. Spin-stabilized spacecraft can take advantage of centrifugal force; three-axis spacecraft or spacecraft with a low spin rate must use one of the following schemes:

Fig. 4.11   Zero-*g* bubble shapes.

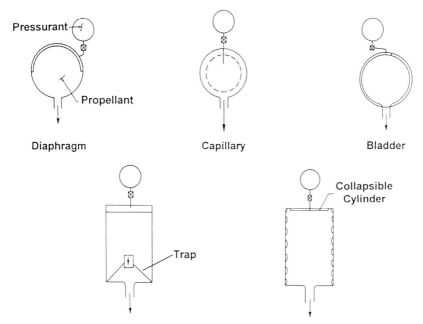

Fig. 4.12   Zero-g propellant control devices.

1) *Capillary devices*, which use surface tension forces to keep gas and liquid separated. These are particularly useful for bipropellant systems like the Space Shuttle and Viking Orbiter because they are compatible with strong oxidizers.

2) *Diaphragms and bladders*, which are physical separation devices made of elastomer or Teflon. These are used by Voyager, Mariner 71, and Magellan. Elastomer types are not compatible with oxidizers.

3) *Bellows*, a metal separation device, used by Minuteman.

4) *Traps*, a check valve protected compartment, used by Transtage.

Figure 4.12 shows the concepts schematically.[6]

Capillary Devices

Capillary devices take advantage of the small pressure differences between a wetting liquid and a gas. For a simple screen device, Fig. 4.13, the ΔP capability is inversely proportional to the radius of the screen openings. The Young–Laplace equation[6] states that the maximum differential pressure across a spherical surface is

$$\Delta P = \frac{2\sigma}{r} \qquad (4.4)$$

where

p = pressure difference sustained by a screen, lb/ft^2
σ = surface tension of the liquid, lb/ft
r = radius of the screen pores, ft

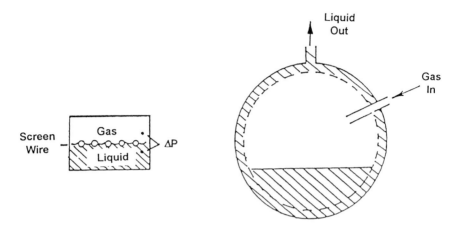

Fig. 4.13   Simple capillary device.

Extremely fine screen will support approximately 2 ft (hydrostatic head) of hydrazine against an adverse 1-g acceleration. The screen devices are shaped to maintain liquid over the tank outlet for any acceleration direction the spacecraft will experience. The capillary device used to control the Space Shuttle reaction control system propellants (N_2O_4 and MMH) is shown in Fig. 4.14. In the Space Shuttle, a system with complicated maneuvering, the capillary system also becomes complex. Both capillary devices and bladders have demonstrated an expulsion efficiency of 99%.

Propellant Control–Spin-Stabilized Spacecraft

A spinning spacecraft can use centrifugal force to control propellant position. Spin rates above about 6 rpm are required for centrifugal force to dominate surface tension and slosh forces[19]; it takes an analysis of the expected spacecraft loads to verify the adequacy of centrifugal force. An expulsion efficiency of over 97% can be expected.[19]

The tank outlet is placed perpendicular to the spin axis, pointing outboard for the liquid to be held over the outlet by centrifugal force. Unfortunately, the tank outlet is then horizontal when the spacecraft is being assembled and tested; in this position, the tank is difficult to drain. Three solutions to this problem are shown in Fig. 4.15, courtesy of Sackheim,[20] along with the pros and cons of each. These alternatives are summarized as follows.

1) The tank can be kept simple and spherical, and the entire spacecraft (or propulsion module) can be rotated to drain the tank. Alternatively, an additional port can be installed in the tank for drainage.

2) An internal drain tube can be installed sloped so that either acceleration vector will drain the liquid.

3) The tank can be shaped so that it will drain with either acceleration vector. Such a tank has a teardrop, or conosphere, shape. This design has been used successfully on a number of recent spacecraft. One disadvantage of this design is that energy dissipation from fuel slosh reduces the stability margin of a dual-spin

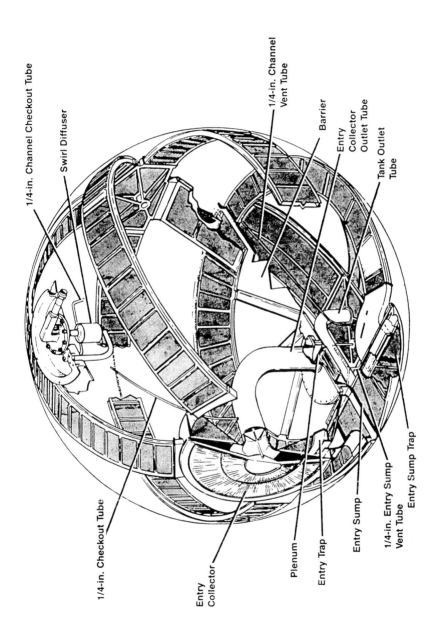

Fig. 4.14   Space Shuttle RCS tank capillary device. (Courtesy Lockheed Martin.)

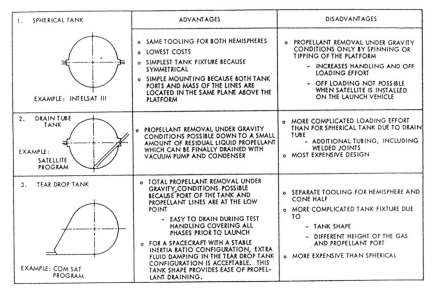

1. SPHERICAL TANK	ADVANTAGES	DISADVANTAGES

EXAMPLE: INTELSAT III	o SAME TOOLING FOR BOTH HEMISPHERES
	o LOWEST COSTS
	o SIMPLEST TANK FIXTURE BECAUSE SYMMETRICAL
	o SIMPLE MOUNTING BECAUSE BOTH TANK PORTS AND MASS OF THE LINES ARE LOCATED IN THE SAME PLANE ABOVE THE PLATFORM
		o PROPELLANT REMOVAL UNDER GRAVITY CONDITIONS ONLY BY SPINNING OR TIPPING OF THE PLATFORM
		  - INCREASES HANDLING AND OFF LOADING EFFORT
		  - OFF LOADING NOT POSSIBLE WHEN SATELLITE IS INSTALLED ON THE LAUNCH VEHICLE

2. DRAIN TUBE TANK
 	
EXAMPLE:
  SATELLITE
  PROGRAM
	o PROPELLANT REMOVAL UNDER GRAVITY CONDITIONS POSSIBLE DOWN TO A SMALL AMOUNT OF RESIDUAL LIQUID PROPELLANT WHICH CAN BE FINALLY DRAINED WITH VACUUM PUMP AND CONDENSER	o MORE COMPLICATED LOADING EFFORT THAN FOR SPHERICAL TANK DUE TO DRAIN TUBE
		  - ADDITIONAL TUBING, INCLUDING WELDED JOINTS
		o MOST EXPENSIVE DESIGN

3. TEAR DROP TANK
 	
EXAMPLE: COM SAT
   PROGRAM
	o TOTAL PROPELLANT REMOVAL UNDER GRAVITY CONDITIONS. POSSIBLE BECAUSE PORT OF THE TANK AND PROPELLANT LINES ARE AT THE LOW POINT
	  - EASY TO DRAIN DURING TEST HANDLING COVERING ALL PHASES PRIOR TO LAUNCH
	o FOR A SPACECRAFT WITH A STABLE INERTIA RATIO CONFIGURATION, EXTRA FLUID DAMPING IN THE TEAR DROP TANK CONFIGURATION IS ACCEPTABLE. THIS TANK SHAPE PROVIDES EASE OF PROPEL-LANT. DRAINING.
		o SEPARATE TOOLING FOR HEMISPHERE AND CONE HALF
		o MORE COMPLICATED TANK FIXTURE DUE TO
		  - TANK SHAPE
		  - DIFFERENT HEIGHT OF THE GAS AND PROPELLANT PORT
		o MORE EXPENSIVE THAN SPHERICAL

Fig. 4.15   Tank configurations for spin-stabilized spacecraft (from Ref. 20).

spacecraft; this reduction is worse than that of a spherical tank and can be as great as a factor of 30.[20]

Propellant Loading

Spacecraft systems use the weight method to measure propellant loaded. The system is moved to a remote area and weighed empty. The propellant is then loaded using special clothing and procedures to protect the personnel. Loading is a hazardous event; one or more dry runs of the procedure precede the actual event. After loading, the system is weighed again. The weight change is the propellant load.

Because of the loading process, it is highly desirable to design any liquid propulsion system to be readily removable from the spacecraft as a module, because:

1) The propulsion system can be taken to the remote area for loading independently of the spacecraft operations.

2) Work can proceed on the spacecraft in parallel with the loading process.

3) If a spill occurs during loading, the equipment at risk is limited.

The loading takes place at the launch site close to the launch date, a time and place when these advantages are very important.

4.4   Pressurization Subsystems

The purpose of a pressurization system is to control the gas pressure in the propellant tanks. For spacecraft systems, the tank pressure must be higher than the engine chamber pressure by an amount equal to the system losses, and a significant delta pressure must be maintained across the injector for combustion stability.

Pressurants

The pressurant, or pressurizing gas, must be inert in the presence of the propellants, and a low molecular weight is desirable. There are two pressurants in use: nitrogen and helium. Although helium provides the lightest system, helium leakage is difficult to prevent; therefore, nitrogen is used if the weight situation will allow it.

Ullage

The gas volume over the propellant is called the *ullage*. This curious term was borrowed from the wine industry.

Blowdown System

Two basic types of pressurization systems are in use today: regulated and blowdown. A blowdown system is shown in Fig. 4.16. In a blowdown system, the tank is pressurized to an initial level, and the pressure is allowed to decay as propellant is used.

The advantages of a blowdown system are: 1) It is the simplest method and, hence, more reliable, and 2) it is less expensive because of fewer components. The disadvantages are: 1) tank pressure, thrust, and propellant flow rate vary as a function of time, and 2) I_{sp} is a second-order function of chamber pressure (for monopropellant hydrazine systems) and drops as a function of time.

Variability of flow rate and engine inlet pressure make the blowdown system difficult to use with bipropellant systems. The disadvantages are slight for monopropellant systems, and modern monopropellant systems use blowdown exclusively.

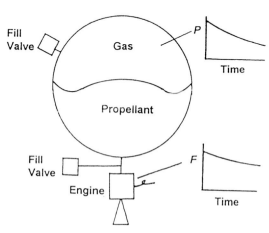

Fig. 4.16 Blowdown pressurization.

Tank Gas Thermodynamics

Blowdown ratio.   The ratio of initial pressure to final pressure is called the
blowdown ratio, B. The blowdown ratio is

$$B = \frac{P_{gi}}{P_{gf}} = \frac{V_{gf}}{V_{gi}} \tag{4.5}$$

and

$$V_{gi} = \frac{V_u}{B - 1} \tag{4.6}$$

$$V_{gi} = \frac{W_u}{\rho(B - 1)} \tag{4.7}$$

where

V_{gi} = initial ullage volume
P_{gi} = initial gas pressure
V_{gf} = final ullage volume
P_{gf} = final gas pressure
V_u = volume of usable propellant
W_u = weight of usable propellant
B = Blowdown ratio

The maximum blowdown ratio is determined by the inlet pressure range the
engines can accept. Ratios of three to four are in common use today; a blowdown
ratio of six has been flown.[20]
  Specific impulse for a monopropellant always decreases with pressure decay
in a blowdown system. The loss is caused by the increase of stay time of the
ammonia in the catalyst bed at the lower pressure. The greater the ammonia stay
time, the greater the ammonia dissociation and the lower the I_{sp}. The I_{sp} loss can
be estimated by

$$I_{sp2} = I_{sp1}(1 - 0.005B) \tag{4.8}$$

where

I_{sp1} = specific impulse at pressure 1
I_{sp2} = specific impulse at pressure 2
B = blowdown ratio, P_1/P_2

For a blowdown ratio of 5, the I_{sp} loss is about 2.5%.
  Equation of state.   From thermodynamics, we know that, for an ideal gas at
any state point, the product of the tank pressure and the ullage volume is

$$PV = WRT \tag{4.9}$$

where

W = gas weight, lb
P = gas pressure, or partial pressure, psf
V = ullage volume, ft^3
R = specific gas constant ($R = 55.16$ for nitrogen; $R = 386.3$ for helium)
T = temperature of the gas, °R

Equation (4.9), the equation of state, relates the thermodynamic parameters for a perfect gas and steady-state conditions. (The departure of real gases from the behavior of perfect gases is normally negligible.) The equation of state is particularly useful for calculating pressurant weight in the form

$$W = \frac{PV}{RT} \tag{4.10}$$

Vapor Pressure

Given time, the gas in the ullage will be an equilibrium mixture of pressurant and propellant vapor. The propellant vapor will saturate the tank gas and reach a partial pressure equal to the propellant vapor pressure. The tank pressure is then the sum of the partial pressure of the pressurant gas and the vapor pressure of the propellant,

$$P_t = P_v + P_g \tag{4.11}$$

The partial pressure of the propellant vapor P_v is the vapor pressure of the propellant at the propellant temperature. The gas pressure P_g is that calculated from the equation of state. Vapor pressure affects tank pressure even if a bladder or a diaphragm is present because the bladder material is permeable. Vapor pressure is usually negligible for fuels but not for oxidizers nor any cryogenics; nitrogen tetroxide has a vapor pressure at room temperature of about 1 atm. Vapor pressures of common propellants are listed in Appendix B.

Example 4.1: Vapor Pressure Effect

Consider a 3.636-ft^3 ullage over a nitrogen tetroxide tank that has been pressurized with 2.547 lb of nitrogen. If the propellant and ullage gas are at 70°F, what is the tank pressure? The nitrogen partial pressure will be, from the equation of state,

$$P_{N_2} = \frac{(2.547)(55.16)(530)}{(3.636)(144)} = 142.21 \text{ psia}$$

However, the tank pressure will be 142.21 plus the vapor pressure of nitrogen tetroxide at 70°F or

$$P_t = 142.21 + 14.78 = 156.99 \text{ psia}$$

Consider this same tank if the temperature is increased to 95°F. The tank volume is 7.570 ft^3. The tank pressure results from four effects:

1) Piston effect of the propellant expansion.
2) Expansion of the nitrogen.
3) Increase in the vapor pressure of the propellant.
4) Expansion of the tank walls (a small effect usually neglected).
The effect of propellant expansion is calculated as follows:

The propellant volume at 70°F is

$$V_p = 7.570 - 3.636 = 3.934 \text{ ft}^3$$

The density of nitrogen tetroxide is (from Appendix B)

$$\rho_{N_2O_4} = 91.06 - (0.0909)(°F - 60)$$
$$\rho_{N_2O_4} = 90.15 \text{ lb/ft}^3 @ 70°F$$
$$\rho_{N_2O_4} = 87.88 \text{ lb/ft}^3 @ 95°F$$

The propellant loaded mass is

$$W_p = 3.934(90.15) = 354.65 \text{ lb}$$

The propellant volume at 95°F is

$$V_p = \frac{354.65}{87.88} = 4.036 \text{ ft}^3$$

The ullage volume at 95°F is

$$V_u = 7.570 - 4.036 = 3.534 \text{ ft}^3$$

The nitrogen partial pressure, from the equation of state, is

$$P_{N_2} = \frac{(2.547)(55.16)(555)}{(3.534)(144)} = 153.22 \text{ psia}$$

The nitrogen tetroxide vapor pressure is 27.5 psia at 95°F (from Appendix B), and the tank pressure is

$$P_t = 153.22 + 27.5 = 180.7 \text{ psia}$$

Temperature Effects.   In equilibrium, the tank gas temperature approximates the propellant temperature because the propellant heat capacity is much greater than that of the gas. Isothermal conditions prevail in a propellant tank when propellant is withdrawn slowly. For an isothermal process,

$$P_1 V_1 = P_2 V_2 \tag{4.12}$$

The tank gas pressure and temperature as functions of time during blowdown can be bounded by assuming isentropic or isothermal conditions. Figure 4.17 shows the two processes for a given amount of propellant drained at a high rate and a slow rate.

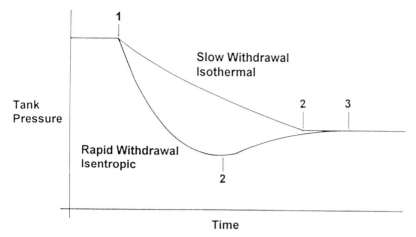

Fig. 4.17   Isothermal and isentropic blowdown.

If the outflow of propellant is slow, heat transfer tó the gas will keep the gas temperature fixed at or near the propellant temperature, and the process will be *isothermal*. If the outflow is rapid, the gas temperature will drop well below the propellant temperature and the process will be *isentropic*. In this case, the gas temperature will rise after the burn, point 2 in Fig. 4.17, and approach the propellant temperature asymptotically with time, point 3 in the figure.

Isentropic expansion.   If the propellant is withdrawn rapidly, the process is isentropic and the pressurant temperature will drop well below the propellant temperature. A long time after a rapid withdrawal, the tank gas will have reached an equilibrium temperature near that of the propellant, and the endpoints will be related isothermally. For an isentropic process,

$$P_1 V_1^k = P_2 V_2^k \tag{4.13}$$

$$\frac{T_2}{T_1} = \left(\frac{V_1}{V_2}\right)^{k-1} \tag{4.14}$$

If isentropic conditions are assumed, the tank pressure at any time during blowdown is

$$P(t) = P_{gi}\left(\frac{V_{gi}}{W_p(t)/\rho + V_{gi}}\right)^k \tag{4.15}$$

Isothermal expansion.   If propellant is withdrawn slowly, the expansion of the ullage gas will be isothermal and will approach the propellant temperature. Isothermal expansion is the most common case. For isothermal conditions,

$$P_2 = P_1 \frac{V_1}{V_2} \tag{4.16}$$

During the blowdown process, the tank pressure at any time t, is

$$P(t) = P_{gi}\left(\frac{V_{gi}}{W_p(t)/\rho + V_{gi}}\right) \qquad (4.17)$$

where

$P_g(t)$ = tank gas pressure at any time t, psia
$W_p(t)$ = total propellant weight removed from the tank in the time
        interval ending at time t, lb
ρ = propellant density, lb/ft^3

Example 4.2: Tank Gas Thermodynamics

The Magellan spacecraft successfully completed its first orbit trim maneuver on May 17, 1991. The purpose of the maneuver was to move the ground track approximately 10 km. The Magellan system is a monopropellant blowdown system using helium pressurant in a single bladder tank. The conditions before the maneuver were:

Spacecraft mass = 2526.72 lb
Propellant mass = 247.42 lb
Helium loaded = 0.571 lb
Tank volume = 6.632 ft^3
Specific impulse = 212.2 s
Tank pressure = 313.6 psia
Helium temperature = 86.0°F
Propellant temperature = 85.1°F

The measured velocity change was 11.20 m/s. Predict the ullage conditions after the burn.

Calculate propellant density and volume. From Appendix B, hydrazine density is

$$\rho = 65.0010 - 0.0298(85.1) - 8.7023E - 6(85.1)^2$$
$$\rho = 62.402 \text{ lb/ft}^3$$
$$V_p = 247.42/62.402 = 3.965 \text{ ft}^3$$

The gas volume is then

$$V_g = 6.632 - 3.965 = 2.667 \text{ ft}^3$$

Check the data for consistency using the equation of state,

$$W_g = \frac{(313.6)(144)(2.667)}{(386.25)(546)} = 0.5711 \text{ lb}$$

This result checks the loaded helium mass that indicates that the data are consistent and that there has been no helium leakage.

The propellant mass to produce an 11.2 m/s (36.745 fps) burn is

$$W_p = (2526.72)\left[1 - \exp\left(-\frac{36.745}{(32.1740)(212.2)}\right)\right] = 13.56 \text{ lb}$$

The spacecraft mass after the burn is 2513.16 lb, and the remaining propellant mass is 233.86 lb.

If isothermal conditions are assumed, the propellant volume is

$$V_p = 233.86/62.402 = 3.748 \text{ ft}^3$$

The gas volume is $6.632 - 3.748 = 2.884 \text{ ft}^3$, and the predicted gas pressure is

$$P_g = 313.6\frac{2.667}{2.884} = 290.00$$

The measured Magellan gas pressure was 292.1 psia.[46]
This example can be worked using the PRO software.

Blowdown system performance. With a blowdown system, the chamber pressure, propellant flow rate, and thrust become functions of tank pressure (and, hence, of time). The flow rate at any time is a function of the hydraulic resistance in the flow path. For incompressible, turbulent flow,

$$\Delta P = R\dot{w}_p^2 \tag{4.18}$$

where

R = resistance coefficient for the propellant flow path, psi-s²/lb²
$\dot{w}_p$ = propellant flow rate, lb/s

Thus, the pressure drop from tank pressure to chamber pressure is

$$P_{gt} - P_c = (R_1 + R_2 + R_3)\dot{w}_p^2 \tag{4.19}$$

where

R_1 = flow coefficient for the propellant lines and fittings from the tank to the thrust chamber valve
R_2 = flow coefficient for the thrust chamber valve
R_3 = flow coefficient for the catalyst bed

The chamber pressure produced by a given propellant flow rate is

$$P_c = \frac{I_{sp}\dot{w}_p}{A_t C_f} \tag{4.20}$$

When the system is operating in quasi-steady-state conditions, Eqs. (4.19) and (4.20) can be solved for P_c, and $\dot{w}_p$. Let R be the flow coefficient from the propellant tank through the catalyst bed,

$$R = R_1 + R_2 + R_3 \tag{4.21}$$

Then, substituting Eq. (4.20) into Eq. (4.19) yields

$$R\dot{w}_p^2 + \frac{I_{sp}}{A_f C_f}\dot{w}_p - P_{gt} = 0 \tag{4.22}$$

Equation (4.22) can be solved as a quadratic to produce

$$\dot{w}_p = \frac{1}{2R}\left[\sqrt{\frac{I_{sp}^2}{A_t^2 C_f^2} + 4R P_{gt}} - \frac{I_{sp}}{A_t C_f}\right] \tag{4.23}$$

Equation (4.23) is the positive root that is the only possible one. To improve computational accuracy, it is desirable to eliminate the subtraction of two numbers of similar size. Using numerical analysis techniques to eliminate the negative sign yields

$$\dot{w}_p = \frac{2P_{gt}}{\sqrt{\dfrac{I_{sp}^2}{A_t^2 C_f^2} + 4R P_{gt}} + \dfrac{I_{sp}}{A_t C_f}} \tag{4.24}$$

For computational convenience, let

$$J = \frac{I_{sp}}{A_t C_f} \tag{4.25}$$

Then, solve

$$\dot{w}_p = \frac{2P_{gt}}{\sqrt{J^2 + 4R P_{gt}} + J} \tag{4.26}$$

As tank pressure drops with time, the propellant flow rate can be obtained from Eq. (4.26), and chamber pressure can be obtained from Eq. (4.20). It is said that an engine takes the square root of flow rate to produce thrust. Equation (4.26) shows why this is true, at least approximately.

The pressure loss across a catalyst bed is designed to be between 30 and 40 psi at maximum flow rate.[15] The thrust chamber valve pressure drop is usually about 50 psi. If the pressure drop is known, the flow coefficient can be computed from

$$R = \frac{\Delta P}{\dot{w}_p} \tag{4.27}$$

The pressure loss from the tank to the thrust chamber valve is system-dependent.

Some spacecraft (notably Magellan and Intelsat V) use a single-shot repressurization in conjunction with blowdown. In the Magellan system, the auxiliary pressurant is stored in a 288-in.[3] titanium pressurant tank at 2300 psi. The auxiliary gas is released into the propellant tank by an ordnance valve. Figure 4.18 shows the thrust vs expended propellant for one 100-lb thruster using the Magellan blowdown-repressurization system.

The advantage of repressurization is that the variation in thrust can be reduced and the loss of I_{sp} can be reduced. In the Magellan case, minimum thrust was raised from about 40 lb to about 60 lb, as shown in Fig. 4.18.

Flight bipropellant systems have not yet used a blowdown pressurization, although there is serious work going on in the area,[22] because of the difficulty in holding the two propellant tanks at the same pressure and the concern about operating bipropellant engines over a wide pressure range.

Regulated Systems

A regulated system controls the pressure in the tank (or tanks) at a preset pressure (see Fig. 4.19). The pressurant (pressurizing gas) is stored at a high pressure (3000–5000 psi). The engine is supplied propellant at a tightly controlled lower pressure, and thrust does not vary during propellant consumption. The penalty for regulation is complexity. For monopropellant systems, the only advantage of regulation is a constant thrust as a function of time; for bipropellant systems, regulation is essential in order to keep the flow rate of each propellant constant and at the correct mixture ratio.

The pressurant tank weight is essentially constant for any initial pressure for a given gas weight; for example, it takes about 7 lb of tank to contain 1 lb of helium. Thus, the initial storage pressure selected is a second-order trade; pressures from 3000 to 5000 psi are used. The regulator requires an inlet pressure around 100 psi higher than the outlet pressure in order to operate properly. Therefore, the unusable gas trapped in the pressurant tank must be computed at the minimum regulator inlet pressure.

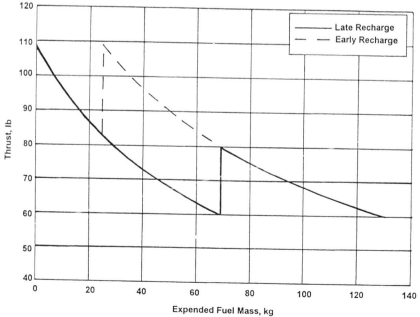

Fig. 4.18   Blowdown with repressurization. (Courtesy Lockheed Martin.)

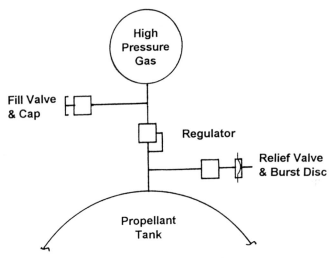

Fig. 4.19   Regulated pressurization system.

The minimum ullage volume must be about 3% of the propellant tank volume for the regulator to have a stable response when outflow starts. In addition to the pressure regulator, relief valves are necessary to protect the tank in case the regulator fails to open. It is good practice to place a burst disk upstream of the relief valve so that pressurant is not lost overboard at the valve seat leakage rate. This precaution is particularly important if helium is the pressurant. For bipropellant systems, each propellant tank ullage must be isolated to prevent vapor mixing.

The pressure balance for a regulated system is a single point since tank pressure does not vary.

4.5 Tankage

The major dry weight component in a liquid propellant propulsion system is the *tankage*. In this section, we will demonstrate how tank weight is calculated. First, the volume of the tank must be set. For a regulated system, the weight of the gas stored is the sum of: 1) the weight of the gas initially loaded into the propellant tank, 2) the weight of the gas transferred into the propellant tank, and 3) the weight of the residual gas left in the pressurant tank when propellant is depleted. For a propellant tank, there are four components to the volume: 1) the initial ullage, 2) the usable propellant volume, 3) the unusable propellant volume, which is 3% to 4% of usable, and 4) the volume occupied by the zero-*g* device.

Example 4.3: Blowdown Propellant Tank Sizing

What is the propellant tank size for a blowdown hydrazine monopropellant system with the following specifications:

$$I_{sp} = 225\text{s}$$
$$I = 10,000 \text{ lb-s}$$
$$B = 4$$

Assume an elastomeric diaphragm, a spherical tank, and a design temperature range of 40°F to 100°F.

The usable propellant weight $W_u = 10,000/225 = 44.4444$ lb. The density of hydrazine at $100° = 61.9378$ lb/ft³ (Appendix B) and the volume of the usable propellant is $V_u = 1239.95$ in.³ The volume of unusable propellant is about 3% of usable for a monopropellant system; thus, the total volume of propellant is $V_p = 1277.15$ in.³ The initial ullage volume is

$$V_{gi} = \frac{1239.95}{3} = 413.32 \text{ in.}^3$$

The tank volume before provisions for the bladder is

$$V_p + V_{gi} = 1277.15 + 413.32 = 1690.47 \text{ in.}^3$$

A diaphragm is essentially a hemispherical shell that nests within the tank shell. The internal radius of the diaphragm is approximately

$$r = \sqrt[3]{\frac{(0.75)(1690.47)}{\pi}} = 7.3899$$

The area of the diaphragm is half the surface of a sphere

$$A_d = 2\pi r^2 = 2\pi (7.3899)^2 = 343.13 \text{ in.}^2$$

The diaphragm thickness can be expected to be about 0.075 in.; therefore, the volume of the diaphragm can be approximated:

$$V_d = 0.075(343.13) = 25.73 \text{ in.}^2$$

and the required volume of the tank is 1716.20 in.³

After the volume required is determined, the tank weight may be calculated. The common tank configurations are spherical tanks and barrel tanks with hemispherical domes.

Spherical Tanks

Tank weights are a by-product of the structural design of the tanks. For spheres, the load in the walls is pressure times area, as shown in Fig. 4.20. The force PA tending to separate the tanks is

$$PA = P\pi r^2 \qquad (4.28)$$

where

P = maximum gauge pressure in the tank, psig (Pressure above atmospheric—gauge and absolute pressures are the same in space).
r = Internal radius of the tank, in.

And, since stress is load divided by the area carrying the load,

$$\sigma = \frac{P\pi r^2}{2\pi r t} = \frac{Pr}{2t} \qquad (4.29)$$

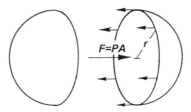

Fig. 4.20   Spherical tank stresses.

and

$$t = \frac{Pr}{2\sigma} \tag{4.30}$$

where

σ = allowable stress, psi
t = thickness of the tank wall, in.

It is difficult to machine to thicknesses less than 0.010 in. and maintain tolerances. For larger tanks with diameters greater than 30 in., a minimum wall of 0.020 to 0.025 in. should be held. If stress considerations allow thinner walls, arbitrarily increase the wall to minimum gauge. (An emerging technology allows thinner minimum gauges by spin-forming the tank hemispheres.)

When the thickness of the walls is known, the weight of the tank membrane can be calculated as follows:

$$R = r + t \tag{4.31}$$
$$W = 4/3\pi\rho(R^3 - r^3) \tag{4.32}$$

where

W = tank membrane weight, lb
ρ = density of the tank material, lb/in.^{3}
R = outside radius of the sphere, in.

Cylindrical Barrels

For cylindrical barrels, the hoop stress is twice that in a spherical shell, shown in Fig. 4.21.

$$t = \frac{pr}{\sigma} \tag{4.33}$$

The weight of the barrel is

$$W = \pi L\rho(R^2 - r^2) \tag{4.34}$$

where L = length of the barrel section, in inches.

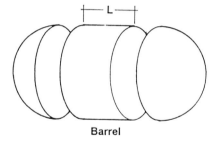

Barrel

Fig. 4.21   Cylindrical tank.

Penetrations and Girth Reinforcements

The membrane weight is for a perfect pressure shell. Areas of reinforcement, called land areas, are needed for 1) girth welds (the weld between two hemispheres), 2) penetration welds (welds for inlet and outlets), 3) bladder attachment (attachment between a bladder and the tank wall), 4) structural mounting of the tank.

Weight must be added to the membrane weight to account for these reinforcements. Weld areas must be reinforced because of the reduction in material properties near welds. For the girth weld, add the weight of two reinforcing bands with a thickness of t, a width of 2 in. each, and a length equal to the circumference of the sphere, as shown in Fig. 4.22. For the penetrations, add a 5-in.-diam disk, centered on the penetration, with a thickness of t. For mounting pads on the tank, add about 2% of the supported weight. It is important to design the tank mounting so that the tank can expand and contract without constraints. Otherwise, expansion will put unexpected loads into the walls. These reinforcements normally add about 25% to the shell weight.

Example 4.4: Spherical Propellant Tank Design

Design a spherical titanium propellant tank with the following specifications:
Maximum working pressure $= 675$ psig
Internal volume $= 1716.18$ in.^{3}
Safety factor of 1.50 on burst
Safety factor of 1.25 on yield
Elastomeric diaphragm
Provide land areas for penetrations at each pole, a girth weld, and a structural mount pad at the lower pole. Calculate the membrane thickness first:

$$r = \sqrt[3]{\frac{0.75(1716.18)}{\pi}} = 7.4272 \text{ in. (inside, minimum)}$$

2 in

t $2t$

Fig. 4.22   Land area for welds.

For titanium forgings, an ultimate strength of 150,000 psi and yield strength of 140,000 psi can be expected. From Eq. (4.32), with allowable stress of 100,000 psi,

$$t = \frac{(675)(7.4272)}{2(100000)} = 0.025 \text{ in.}$$

Regardless of strength requirements, a minimum thickness of 0.01 in. should be maintained to avoid handling and assembly damage; these are called *minimum guage considerations*. A machining tolerance of ±0.001 can be maintained in machined forgings. The thickness 0.025 is the minimum acceptable; therefore,

$$t = 0.026 \pm 0.001 \text{ in.}$$

Weight estimates should be made for the maximum thickness. With the thickness determined, the tank outside radius is calculated:

$$R = 7.4272 + 0.027 = 7.4542 \text{ in. (maximum)}$$

Calculate membrane weight using a density of 0.16004 lb/in.³ (see Appendix B):

$$W = \tfrac{4}{3}\pi(0.16004)[(7.4542)^3 - (7.4272)^3] = 3.01 \text{ lb}$$

Now, calculate the reinforced areas, starting with the girth weld land,

$$W = 2\pi R t w \rho$$
$$W = 2\pi(7.4542)(0.027)(4)(0.16004) = 0.810 \text{ lb}$$

Two penetrations are required, one at each pole. The reinforcing land around each penetration can be assumed to be a 5-in.-diam disk of thickness t. The weight of these reinforcements is

$$W = 2\pi r^2 t \rho = \pi(12.5)(0.027)(0.16004) = 0.170 \text{ lb}$$

The structural attachment weight = 2% of supported weight. The supported weight is:

Membrane	3.01 lb
Girth land	0.81
Penetrations	0.17
	3.99 lb
Structural attachment	0.08
Tank shell weight	4.07 lb

Now, calculate the diaphragm volume and weight, assuming a thickness of 0.075 in. and a density of 0.03 lb/in.³:

$$V_d = \tfrac{2}{3}\pi[(7.4272)^3 - (7.3522)^3] = 25.734 \text{ in.}^3$$
$$\text{Diaphragm weight} = (0.036)(25.734) = 0.926 \text{ lb}$$

A mounting bead runs around the edge of a diaphragm for tank attachment. If the bead is estimated to be 1 in. wide,

$$\text{Bead weight} = 2\pi(7.4272)(0.075)(1)(0.036) = 0.1260 \text{ lb}$$

The diaphragm total weight is $0.926 + 0.1260 = 1.05$ lb, and the total volume is 29.17 in.^{3} The tank assembly weight is:

Tank shell	4.07 lb
Diaphragm	1.05
Total	5.12 lb

The available propellant volume is $1716.18 - 29.17 = 1687.01$ in.^{3}.

The procedure used in this example is identical to that used by the PRO software; use the software and compare the results.

Dual Tanks

There is a trend toward dual, identical tanks, especially in communication satellites. The advantages of dual tanks are:

1) Bladder or diaphragm diameter is reduced.

2) The size of tank hemisphere forgings is reduced.

3) There is more flexibility in tank location; with a single tank, balance dictates a location at the center of the spacecraft.

4) Valving can be designed so that tank leakage is not a critical failure. (With modern tank design and test techniques, this is a secondary issue.)

The bladders, diaphragms, and forgings become significantly more expensive above about 30 in. in diameter. For diameters exceeding 30 in., you should consider dual tanks or barrel tanks. (See the Intelsat V design in Sec. 4.7.)

The disadvantage of dual tanks is a small weight increase for the additional tank. Using the PRO software to design single and dual tanks for the following specifications:

Volume, 6.65 ft^3

Elastomer diaphragms

Allowable stress, 100,000 psi

Maximum operating pressure, 575 psi

yields:

	Single tank	Dual tanks
Tank diameter (inside)	27.998 in.	22.222 in.
Tank shell weight	19.74 lb	20.78 lb (2)
Diaphragm weight	3.54 lb	4.54 lb (2)
Tank assembly	23.28 lb	25.32 lb (2)

Materials

Essentially, the only tank material in use is titanium, which has high strength and light weight and is compatible with hydrazine, monomethyl hydrazine, and nitrogen tetroxide. Steel and aluminum are used, but rarely.

Propellant and pressurization lines and fittings are typically stainless steel. Special vacuum arc remelt stainless steel varieties are required for the fittings in order to prevent leakage through flaws in the base metal. Modern systems are all welded to prevent leakage. A fusion-bonded bimetallic fitting is used between the titanium tanks and the stainless steel lines.

4.6 Monopropellant System Design Example

This section illustrates how the information in the previous sections can be used to produce a complete monopropellant system design to the phase A level. The spacecraft under consideration is a geosynchronous communication satellite that is placed in orbit by Atlas Centaur and an apogee kick motor. The spacecraft is three axis–stabilized except during the solid motor burn, for which the spacecraft

Table 4.8   Monopropellant propulsion system design process

1) Define the requirements: Requirement	From
Steady-state impulse required (Set maneuvering thrust level for $T/W>1$)	Mission design
Pulsing impulse required	ACS
Maneuvering thrust level required	ACS
Wheel unloading, if any	ACS
Minimum rpm of spinner	ACS
Maximum moment arm	General arrangement
Fault protection	Customer
Temperature limits	Thermal control

2) Calculate the propellant required; add margin.
3) Select the propellant control device.
4) Decide dual vs single propellant tanks.
5) Decide propellant tank type, sphere, barrel, conosphere.
6) Select the pressurant: helium if mass is critical; otherwise, nitrogen
7) Select pressurization system type, and set the performance parameters, maximum tank pressure, and blowdown ratio.
8) Design the propellant tank.
9) Design the engine modules and general arrangement.
10) Design the system schematic; plan redundancy.
11) Calculate system mass.
12) Conduct trade studies of system alternatives; repeat steps 1–12.

is spun and despun. The steps in a monopropellant propulsion system design are summarized in Table 4.8.

Requirements

The following requirements have been established for the system:
Steady-state impulse required:
N/S stationkeeping 52,800 lb-s
E/W stationkeeping 5720 lb-s
Reaction wheel unloading 8510 lb-s
Pulsing impulse required:
Spin/despin/nutation damping 2100 lb-s
Initial orientation 3500 lb-s
Thrust level required:
1) Minimum impulse bit required in roll; select 0.1-lb thrusters
2) Pitch and yaw thrusters are also used for stationkeeping; thus a compromise is required between minimum impulse bit and maneuver time minimization. Select 5.0-lb thrusters
Maximum moment arm: Atlas medium fairing allows 57.5-in. radius maximum[13]
Propellant system temperature range: 40° to 120°F
No single, nonstructural failure can cause loss of more than 50% of the mission objectives or mission duration.

Propellant Inventory

All of the burns are pulsing or short steady-state burns. Select catalyst bed heating to improve both I_{sp} and catalyst bed life. Estimate the average I_{sp} for short burns to be 215 s and for pulsing 110 s. With these assumptions, the propellant inventory, Table 4.9, can be constructed.

Providing propellant reserves is always an area for lively project discussion; alternatively, reserves may be specified by your customer. A 50% reserve is not excessive for early stages of a project.

Table 4.9   Propellant Inventory

1) Propellant for short burns	673030/215 = 311.8 lb
2) Propellant for pulsing	5600/110 = 50.9
3) Reserves	50% = 180.0
4) Subtotal: usable propellant	= 542.7
5) Trapped propellant,	30% of usable = 16.3
6) Loading uncertainty	0.5% of usable = 2.7
7) Loaded propellant	561.7 lb

Initial Selections

Now, let us make some arbitrary choices so that the design can proceed; these choices can be revisited by later trade studies. Initial choices are:

Blowdown pressurization (A safe choice; blowdown is almost universally used for monopropellant systems.)

Blowdown ratio = 4

No repressurization (Repressurization should be the subject of a later trade study.)

Helium pressurant

Initial pressure = 525 psi

Diaphragm propellant control device

Dual spherical tanks (A rapid calculation will show that this propellant load is too big for a single tank.)

Allowable stress in titanium = 100,000 psi

Propellant Tank Design

The propellant tank volume should be calculated based on hydrazine density at maximum temperature, 120°F; from Appendix B, the density is 61.3043 lb/ft³. The hydrazine volume is

$$V_p = \frac{561.7}{61.3043} = 9.1625 \text{ ft}^3 (15{,}832.78 \text{ in.}^3)$$

The volume of the usable propellant is

$$V_u = \frac{542.7}{61.3043} = 8.8525 \text{ ft}^3 (15{,}297.22 \text{ in.}^3)$$

The initial ullage volume is

$$V_{gi} = \frac{15297.22}{3} = 5099.1 \text{ in.}^3$$

The tank volume, before provisions for a diaphragm, is

$$V_t = 5099.1 + 15832.78 = 20931.9 \text{ in.}^3 \text{ (two tanks, 10,466 in.}^3\text{each)}$$

Estimating the diaphragm volume at 1% of the total, or 209 in.³, produces a total volume of 21,141 in.³, or a volume of 10,570 in.³, for each tank. The tank can now be designed using the procedure demonstrated in Example 4.4, with the following result:

Tank design:

Inside diameter = 27.229 in.

Wall thickness = 0.0367 ± 0.001 in.

Tank weight = 16.744 lb

Diaphragm Design:

Diaphragm volume = 93.28 in.³

Weight = 3.36 lb

The tank assembly weight is 20.10 lb each, or 40.20 for the pair. The allowance made for the diaphragm is 11 in.³ larger than required. The tank design could be iterated, but the gain would be small.

Pressurant Weight

The adjusted initial ullage volume for each tank is

$$V_{gi} = 10570 - 15832.8/2 - 93.28 = 2560.3 \text{ in.}^3$$

The helium mass loaded in each tank, from the equation of state, is

$$W_g = \frac{(525)(2560.3)}{(12)(386.3)(500)} = 0.580 \text{ lb (1.160 lb total)}$$

Thruster Arrangement

This spacecraft has a spin mode and a three axis–stabilized mode; the latter mode governs thruster arrangement. As shown in Chapter 3, it takes 12 thrusters to apply pure couples to a three axis–stabilized vehicle.

System Schematic

Fault tolerance requirements dictate redundant branches of 12 engines each to accommodate an engine failure. If the thrust chamber valves are placed in series, their dominant failure mode (leakage or failure open) can be tolerated. Thrust chamber valve failure closed can be tolerated because the engines are redundant. (Series-parallel valves are an alternative solution.) If the dual tanks are cross-strapped, a failure of either propellant system can be accommodated. If ordnance

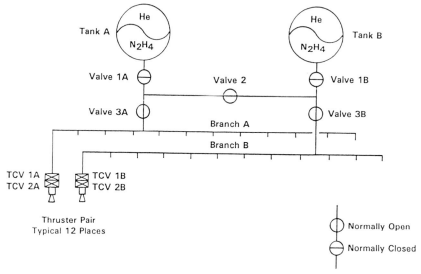

Fig. 4.23   System functional schematic (partial).

Table 4.10   Failure modes and effects

Failure	Corrective action	Mission effect
Thruster A failure-off	Operate with thruster B	None
Thruster A failure-on	Close TCV A1 and A2	None
TCV 1A fail-open	Operate with TCV 2A	None
TCV 1A fail-closed	Operate with thruster B	None
Leakage Branch A below valve 3A	Close value 3A	None
Leakage Branch A above valve 3A	Close value 2	Reduced duration
Diaphragm rupture	Close value 2 and 3A	
Valve 1A fail-closed	None; use Branch B	

valves are used for cross-strapping, the system can be sealed from propellant loading to flight use. The resultant, still incomplete, schematic is shown in Fig. 4.23.

Table 4.10 summarizes the logic used to establish the system schematic in a failure modes and effects analysis (FMEA) table. (Some form of FMEA is normally a contractual requirement.) Valves 1A and 1B are opened at the start of the flight; therefore, they should be considered open as you read Table 4.10. Failure types are considered only once; symmetrical failures are not tabulated. Table 4.10 shows that the system in Fig. 4.23 is satisfactorily failure-tolerant. The schematic can now be completed with the addition of: 1) fill and drain provisions, 2) heaters and thermostats, 3) instrumentation, and 4) filters. The schematic with these additions is shown in Fig. 4.24.

Fill and drain connections are needed to load helium and hydrazine. The fill valves must be capped to provide a double seal against leakage.

Hydrazine must be prevented from freezing; the freezing point is about 2°C. If freezing should occur, the hydrazine shrinks. Line rupture will occur during thaw if liquid fills the volume behind the frozen hydrazine and is trapped. Satellite failures have occurred as a result of this process. The power required for propulsion heating is a product of the thermal design of the vehicle. The most common technique is to provide heaters and thermostats on the lines, tanks, and thrust chamber valves. Catalyst beds are also heated to increase performance and bed life.

Instrumentation is selected so that any failure can be detected from analysis of the data. This requires temperature and pressure in each segment of the system. The FMEA should be reviewed with the instrument list in mind. The instrumentation shown in Fig. 4.24 is adequate if ground testing with more complete instrumentation has been performed. Chamber pressure is a valuable measurement; however, the measurement is not usually attempted in very small thrusters. Chamber pressure can be calculated from upstream pressure, given adequate ground-test data.

Filters are required on the ground equipment side of each fill and drain connection, and so additional flight filters are not needed. Filters are required downstream of all pyro valves and upstream of all valve seats. For the schematic shown in

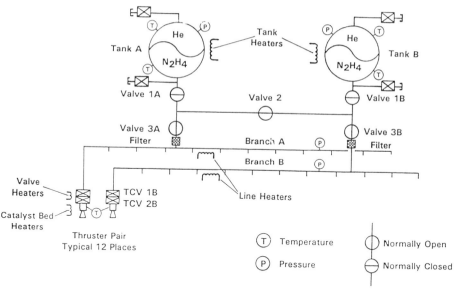

Fig. 4.24   System functional schematic.

Fig. 4.24, two filters are adequate. Several filters that have a 15- to 20-μ absolute rating are manufactured for this use.

First-Pulse Considerations

The first pulses from a flight system are usually required to stabilize the residual rotation rates just after separation from the upper stage. This process is very important; the first pulses must have the full impulse that the attitude-control system expects. For good first pulses, the liquid system must 1) be gas-free, 2) be filled with hydrazine from the tank down to the thrust chamber valves, and 3) have a hot catalyst bed.

Turning the catalyst bed heaters on before separation is easy; filling the hydrazine system is not. At launch, the hydrazine system is full of dry nitrogen from valves 1A and 1B down to the thrust chamber valves. If nothing were done, the first pulses would be nitrogen, not hot gas. One solution to this problem is to open valves 1A and 1B during the upper-stage flight and then open all thrust chamber valves until each chamber temperature rises. This approach may not be satisfactory to the launch vehicle because of the upsetting torques produced and because of exhaust impingement on the upper stage. Another approach is to open the thrust chamber valves during the upper-stage flight with valves 1A and 1B closed. This step bleeds the trapped dry nitrogen into the hard vacuum of space. If valves 1A and 1B are now opened, hydrazine would fill the lines at very high velocity. When the hydrazine reaches the thrust chamber valves, the deceleration shock can be energetic enough to decompose the hydrazine. To avoid this problem, orifices are required in the propellant lines to limit the initial hydrazine flow rate. If these orifices can be made small enough to force orderly initial filling and large enough not to obstruct the maximum thruster demand, the problem is solved. If not, separate bleed lines are required for the orifice flow; this is additional hardware not shown in Fig. 4.24. A third alternative is to launch with the hydrazine filled to the thrust chamber valves. This alternative requires demonstration of its safety to all parties, including the launch vehicle team.

System Mass Estimate

The system is now well enough defined for the first mass estimate shown in Table 4.11.

The thruster weights were taken from Fig. 4.7 at 0.4 lb; however, Fig. 4.7 was compiled for thruster and one valve. The mass was adjusted upward to account for the additional valve. The remaining masses in Table 4.11 were estimated by analogy or calculated in the previous paragraphs.

Trade Studies

The first cycle of propulsion system design has been completed and a baseline established. The next step is to find ways to improve it. Potential improvements should be evaluated by comparison to the baseline by way of trade studies. Some of the trade studies that could be made against the above baseline are:

Table 4.11   Propulsion mass estimate

Components	Unit, mass, lb	No.	Total, mass, lb
Propellant:			561.7
Usable	271.30	2	542.7
Unusable	9.5	2	19.0
Pressurant:	0.6	2	1.2
Feed system:			
Tanks	16.7	2	33.4
Diaphragms	3.4	2	6.8
Valves	1.3	9	11.7
Filters	0.3	2	0.6
Lines and fittings	10.0	—	10.0
Temp. transducers	0.1	28	2.8
Pres. transducers	0.3	4	1.2
Heaters	2.0	—	2.0
Thrusters and valves			
5.0 lb	1.0	16	16.0
0.1 lb	0.7	8	5.6
Wet mass			653.0
Burnout mass			110.3
Dry mass			90.1

1) Evaluate the use of repressurization.
2) Evaluate the use of nitrogen pressurant.
3) Evaluate electrothermal thrusters for the N/S burns.

4.7 Flight Monopropellant Systems

In this section, several monopropellant systems with a successful flight history are described. They are arranged in the order of their launch date and hence in the order of increasing design sophistication. Each made a significant contribution to monopropellant system technology. The characteristics of the systems are summarized in Table 4.12.

Landsat 3

The Landsat 3 offers an illustration of a flight monopropellant system at its simplest. Ground control uses the system to establish precise orbital parameters after orbit insertion and to make orbit adjustments throughout the mission in order to maintain overlapping coverage in the imagery.

The system, shown in Fig. 4.25, is constructed as a single module consisting of three rocket engines, a propellant tank, and a feed system.

Each of the engine assemblies consists of a series-redundant propellant valve, a catalyst bed, and a nozzle. Operation of the solenoid valves by electrical command produces thrust. Through a propellant fill valve, 67 lb of anhydrous hydrazine are loaded. The single spherical titanium propellant tank contains an elastomeric diaphragm for propellant position control. The nitrogen pressurant is loaded through a fill valve. The pressurization is by a 3.3 to 1 blowdown system. The thrust range operating in a blowdown mode is from 1.0 lb initially to 0.3 lb finally. The system total impulse is 15,000 lb-s.

The system is mounted to the spacecraft sensory ring, with the three thrusters located along the pitch and roll axes. The thrusters are aligned such that each thrust vector passes approximately through the spacecraft center of mass. With

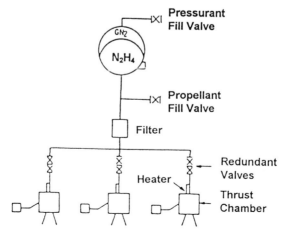

Fig. 4.25   Landsat 3 propulsion system (from Ref. 24, pp. 3–19).

Table 4.12 Flight monopropellant systems

	Mariner 4	Landsat	Viking	HEAO	Voyager	Pioneer Venus	Intelsat V	IUS	Magellan
Launch date	1964	1972	1976	1977	1977	1978	1980	1982	1989
Attitude control	3 Axis	3 Axis	3 Axis	3 Axis	3 Axis	Spin	3 Axis	3 Axis	3 Axis
No. thrusters	1	3	4, 3	12	16, 4, 4	7	20	12	12, 4, 8
Initial thrust, lb	50	1.0	10, 600	1.1	0.2, 5, 100	1.5	0.1, 0.6, 5		0.2, 5, 100
Pressurization	Regulated	Blowdown	Blowdown	Blowdown	Blowdown	Blowdown	Blowdown	Blowdown	Blowdown
Pressurant	N_2	N_2	N_2	N_2	N_2	He	N_2	N_2	He
No. prop tanks	1	1	2	2	1	2	2	1, 2, or 3	1
Initial pressure, psia			530	350	450	350	270		450
Blowdown ratio	—	3.3		3.5		1.8	1.8		4
Repressurization	—	No	No	No	No	No	Yes	No	Yes
Propellant Control	Bladder	Diaphragm	Deceleration	Diaphragm	Diaphragm	5 rpm spin	Capillary	Diaphragm	Diaphragm
Tank shape	Spherical	Spherical	Spherical	Spherical	Spherical	Conoshpere	Barrel	Spherical	Spherical
Crossover	—	—	—	Yes	—	Yes	Yes	Yes	No
Dry mass, lb	26.7	67		56.2			78	123/Tank	135
Propellant mass	21.5		185	300	230	86.2	410		293.2
Features	Slug starts	Simplicity	Throttlable		400,000 cycle pulsing		Electro-thermal thrusters	Removable tanks	
Primary Reference	23	24	25	30	16	27	28	26	29

these thrust vectors, the system is capable of imparting incremental velocity to the spacecraft to correct orbital errors and perturbations.

More complicated monopropellant systems would provide attitude control pulsing, thrust vector control, and maneuvering. These capabilities would be provided by additional thrusters. A planetary system would require additional redundancy, and the system would be sealed by ordnance valves to prevent leakage during coast periods.

The highly successful Landsat series of spacecraft was designed by Lockheed Martin (then General Electric) under contract to NASA Goddard Space Flight Center.

High-Energy Astronomy Observatory System

The purpose of the three high-energy astronomy observatory (HEAO) spacecraft was to carry three different astronomy observatories into low Earth orbit. The scientific payloads came from Europe as well as the United States. The HEAO monopropellant blowdown system provided attitude-control torques with six thruster modules. The system schematic is shown in Fig. 4.26.

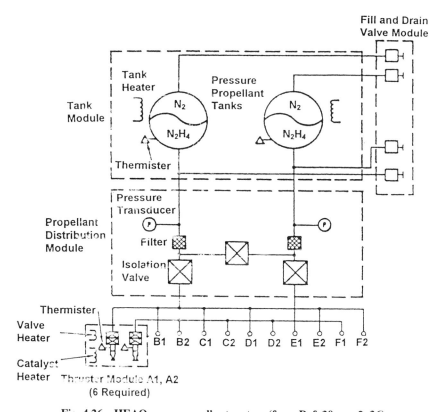

Fig. 4.26   HEAO monopropellant system (from Ref. 30, pp. 2–36).

Propellant is contained in two titanium tanks with crossover capability. The nominal propellant load was 300 lb. Maximum initial tank pressure was 350 psia; the blowdown ratio was 3.5. The system is divided into two redundant halves so that any leakage failure can be isolated and the remaining tank used fully. The twelve identical thrusters have a beginning-of-life thrust level of 1.1 lb and, after blowdown, a thrust level of 0.3 lb. System joints are welded or brazed.[30]

The HEAO spacecraft was designed by TRW for NASA. Three observatories were launched; all were successful.

Viking Lander

The Viking spacecraft consisted of two vehicles: an Orbiter, which provided Earth–Mars transport and orbital observation of the planet, and a lander, which descended to the surface of the planet. Two Viking spacecraft landed on Mars in the summer of 1976. The objective of the mission was to increase scientific knowledge of the planet and, in particular, to explore the possibility of life on Mars.

The lander had two separate monopropellant propulsion systems. One provided attitude control after separation from the Orbiter and the other provided postaerodynamic terminal deceleration for the landing. The attitude-control system was composed of twelve 8-lb engines with redundant valves, which were fed 187 lb of superpure hydrazine from two 22.14-in.-diam bladder tanks. The pressurant was a nitrogen blowdown system with initial pressure of 348 psia. The attitude-control system was attached to the aeroshell, which was jettisoned when the aerodynamic braking phase of the mission was completed.

The landing system, Fig. 4.27, was composed of three throttlable descent engines, four 10-lb roll engines, and two propellant tanks. The propellant tanks were titanium spheres with an inside diameter of 23.5 in.; diaphragms were not required since aerodynamic deceleration held the propellant over the outlet. The nitrogen blowdown system had an initial pressure of 530 psia; the propellant load was 185 lb. The roll-control engines and valving were identical to attitude-control engines; their higher initial thrust was due to higher tank pressure.

The throttlable descent engines were unique; see Fig. 4.28.[25] The three engines were spaced equally around the lander body. They used the spontaneous Shell 405 catalyst; the thrust at full throttle and initial tank pressure was 600 lb. The variable thrust of the descent engine was provided by a motor-driven throttle valve that controlled the propellant flow rate; valve position was telemetered. The exhaust of the engine was separated into 18 small nozzles to minimize the alteration of the landing site. Pitch and yaw control were provided by throttling. The system was single-string; no redundant engines or valves were provided.

Voyager

The Voyager spacecraft, developed by the Jet Propulsion Laboratory, is one of the most successful scientific spacecraft ever to fly. It was launched in 1979; it gathered scientific data at each of the outer planets arriving at Neptune in August of 1989. It used a monopropellant hydrazine system for orbit trims and attitude control and a solid rocket motor to supplement the launch vehicle energy at launch. The monopropellant system supplied thrust vector control during the solid motor

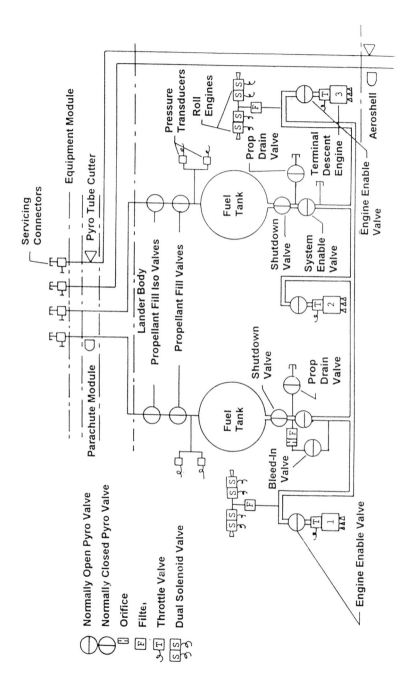

Fig. 4.27 Viking Lander terminal descent system. (Courtesy Lockheed Martin.)

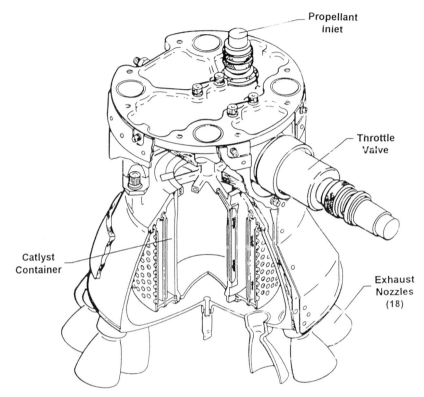

Propellant
Inlet

Throttle
Valve

Catlyst
Container

Exhaust
Nozzles
(18)

Fig. 4.28   Viking terminal descent engine (from Ref. 25, p. 132).

burn. The system schematic is shown in Fig. 4.29 During solid motor burn, pitch and yaw were provided by four 100-lb thrusters; roll control was provided by four 5-lb thrusters. At the completion of the solid motor burn, the spent motor and the eight hydrazine thrusters associated with it were jettisoned. Note the hydrazine disconnect in Fig. 4.29. It is unusual to break a hydrazine system in flight; however, this disconnect operated satisfactorily.

The trajectory correction and attitude-control system contained 230 lb of hydrazine in a 28-in.-diam titanium tank that incorporated an elastomeric bladder. The tank was pressurized initially to 450 psia at 105°F. Pressurization was by blowdown.

Two branches of thrusters were provided, along with the ability to switch pitch/yaw thrusters and roll thrusters independently. Switching was done with latch valves; latch valve position was telemetered. There were sixteen 0.2-lb thrusters and valves. Heaters (1.4-W) were provided for each catalyst bed to maintain a minimum temperature of $240 \pm 15°F$.

During development of the system, a degradation in peak P_c and an increase in ignition delay were observed in vacuum testing the 0.2-lb engines at 10-ms pulse widths. It was determined that the degradation was caused by trace amounts (<0.5%) of aniline in the hydrazine. The use of superpure (Viking

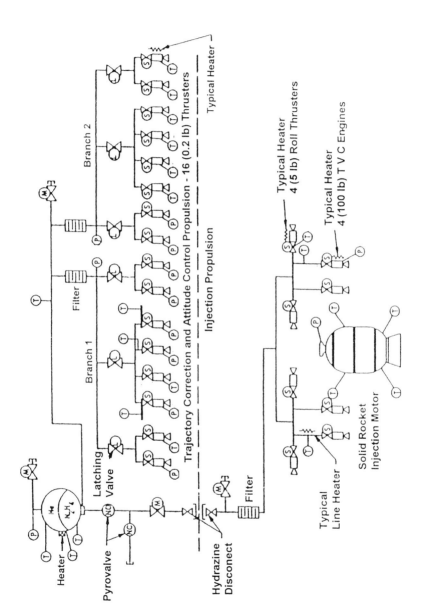

Fig. 4.29 Voyager propulsion system (from Ref. 16).

Table 4.13 Operating cycles of 0.2-lb thruster. Voyager 2 data for 569 days from launch to March, 12, 1979 (from Ref. 16)

Thruster	Voyager 1		Voyager 2	
	Br 1	Br 2	Br 1	Br 2
+Pitch	29,510	521	16,416	2,986
−Pitch	44,235	332	22,612	6,162
+Yaw	44,935	547	22,511	19,027
−Yaw	42,698	485	21,639	9,074
+Roll	50,428	5,777	42,851	0
−Roll	45,236	6,188	44,840	0

Grade) hydrazine cured the problem.[42] The purification technique was developed by Lockheed Martin for the Viking Lander.

The unusually long mission provided interesting flight data on the propulsion system. The cycles accumulated on the thrusters in the first 18 months of the mission are shown in Table 4.13. The total mission time for Voyager 2 from launch to Neptune encounter was 4388 mission days. Linear extrapolation would place the total number of cycles on the Voyager 2 thrusters between 175,000 and 345,000, depending on the sharing between branches 1 and 2. The thrusters were qualified for 400,000 cycles.

The minimum impulse bit of the 0.2-lb thrusters was delivered at 10 ms ± 2 ms. The average 10-ms impulse bit for the first 18 months for the branch 1 thrusters is listed in Table 4.14. The 10-ms pulsing specific impulse measured in flight and ground tests was 110 s.

IUS Reaction-Control System

The IUS reaction-control system provides roll control during SRM burns, stability during coast periods, impulse for accurate orbit adjustment, and collision avoidance after spacecraft separation. The system, shown in Fig. 4.30, uses one, two, or three blowdown propellant tanks, each containing 122.5 lb of hydrazine. The number of tanks, usually two, can be varied to suit the mission. Each tank uses an elastomeric diaphragm to control propellant position.

Table 4.14 Average 10-ms impulse bit (from Ref. 16)

Thruster	Voyager 1	Voyager 2
+Pitch	0.0133	0.0184
−Pitch	0.0156	0.0156
+Yaw	0.0142	0.0156
−Yaw	0.0142	0.0142
+Roll	0.0133	0.0173
−Roll	0.0156	0.0147

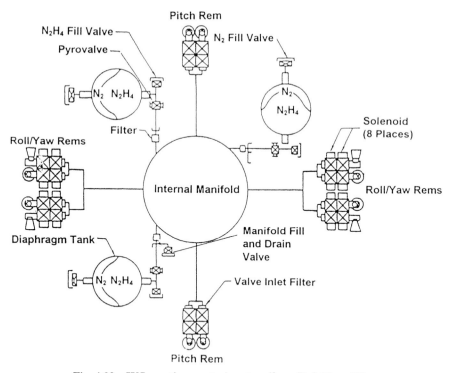

Fig. 4.30   IUS reaction-control system (from Ref. 26, p. 25).

Six two-thruster REMs are used; each thruster has series-redundant valves. The placement of the twelve thrusters provides redundancy for each of the six steering axes. In order to eliminate contamination problems, no thruster points forward toward the payload.

Intelsat V System

The Intelsat V spacecraft is the fifth high-capacity communications spacecraft to be developed for INTELSAT corporation. At the time of its launch, it was the largest communications spacecraft ever built.[31] The Intelsat V monopropellant system consists of two titanium propellant tanks, redundant sets of 10 thrusters each, and associated valves. Figure 4.31 shows the system arrangement.[28]

For orientation and orbit trim, 5-lb thrusters are used. Spacecraft spin/despin, east/west stationkeeping, and pitch/yaw control are performed by 0.6-lb thrusters. Electrothermal thrusters are used for north/south stationkeeping; these 0.1-lb thrusters are backed up by the 0.6 pounders. Roll maneuvers are performed by 0.1-lb thrusters. The plumbing is arranged so that, using isolation valves, either tank can feed either or both sets of thrusters. Capillary propellant management devices control the hydrazine under zero-g or 1-g conditions in all tank positions. The tank internal volume is 140.7 liters and allows a loading of 213 kg of propellant. Nominally, 185 kg are required for the mission. A blowdown pressurization system is used. The tanks are high-strength titanium. All other components are

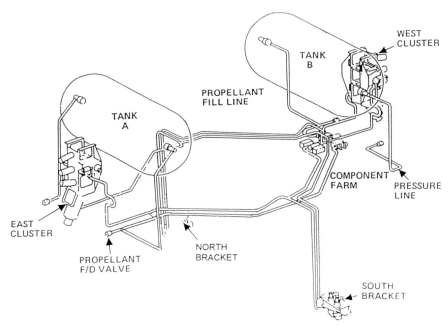

Fig. 4.31   Intelsat V monopropellant system (from Ref. 28).

stainless steel; all joints are welded. Electrochemical 0.07-lb thrusters were chosen for the high I_{sp} (average 304 s) derived from operating at 4000°F. The system schematic is shown in Fig. 4.32.[28]

Intelsat V was developed by Ford Aerospace and Electronics Corp. (now Loral), under contract to INTELSAT. The electrochemical thrusters were developed by TRW Defense and Space Systems Group.[17]

Magellan System

The mission of the Magellan spacecraft provided the first full-surface map of Venus. The map was obtained from an elliptical Venus orbit using a synthetic aperture radar. The Magellan monopropellant blowdown system provided tip-off control during vehicle separations, reaction wheel desaturation, orbit trim maneuvers (for both the sun and Venus orbits), rate damping after any burn, and thrust vector control and attitude control during the solid rocket motor firing. The system was designed so that no single malfunction could prevent mission completion.

Superpure hydrazine was fed from the titanium diaphragm tank through the isolation valves to the four engine modules. Each module contains two 0.2-lb, one 5-lb, and two 100-lb thrusters. The primary purpose of the aft-facing 100-lb engines was thrust vector control during solid rocket motor firing; they were also used for orbit trims. The 5-lb engines and 0.2-lb engines were used for attitude-control functions. The thrusters were redundant and were fed by dual propellant feed systems. Thrusters also serve as a backup in the event of a reaction wheel

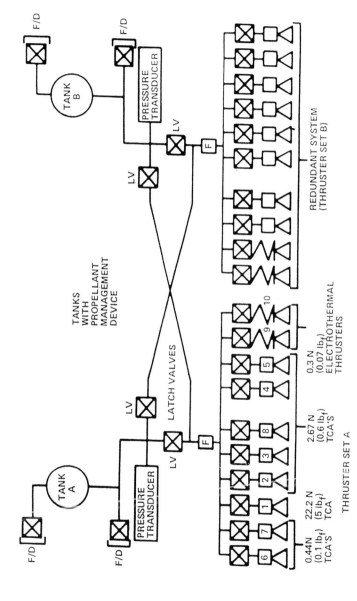

Fig. 4.32 Intelsat V monopropellant system schematic (from Ref. 28).

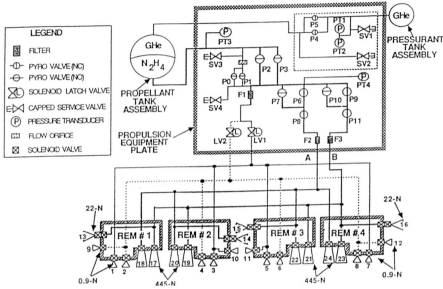

Fig. 4.33   Magellan monopropellant system. (Courtesy Martin Marietta.)

failure. The system schematic is shown in Fig. 4.33.[22] The pressurization system is a blowdown type using helium pressurant. The maximum operating propellant tank pressure is 450 psig and the blowdown ratio is 4. The pressurization system is unusual in that a single recharge is provided by an auxiliary tank containing 288 in.[3] of helium at 3200 psig.

The propellant lines were evacuated during the launch process by opening the thrust chamber valves. During the IUS burn, after deployment from the Space Shuttle, the isolation valves ($P0$ and $P1$ in Fig. 4.33) were opened to allow propellant to fill the lines. Flow rate during line filling was controlled by orifices to prevent shock to the propellant. The propellant and pressurization systems are all welded stainless steel (except for the titanium tanks). There are no mechanical joints in the system.

The liquid propulsion module is readily separable from the spacecraft so that propellant can be loaded at a remote area apart from the spacecraft and off the critical path.

The spacecraft was developed by Lockheed Martin (then Martin Marietta) under contract to the Jet Propulsion Laboratory. The spacecraft was launched on Atlantis, May 9, 1989. The Venus radar mapping mission was completely successful. Mission objectives were expanded to include aerobraking and gravity measurements at low altitude.

Problems

4.1   Prepare a propellant inventory for a monopropellant system to meet the following requirements:

1) Translational maneuver $\Delta V = 200$ m/s at an average I_{sp} of 225 s.
2) Spacecraft wet weight = 1100 lb.
3) Attitude-control total impulse = 15,900 lb-s in pulse mode at an average I_{sp} of 130 s.
4) Propellant reserve = 35% of usable.

4.2 What is the weight of nitrogen remaining in a 10.2-ft³ propellant tank if it is assumed that all of the propellant is consumed, the tank pressure at burnout is 350 psi, and the nitrogen temperature is 60°F?

4.3 How much does a titanium pressurant sphere weigh if the allowable stress is 100,000 psi, the internal volume is 0.136 ft³, and the maximum operating pressure is 3600 psi? What weight of helium does it contain at 3000 psia and 70°F? (This is the Magellan repressurization sphere. The actual tank weight was 3.32 lb, and it held 0.278 lb of helium. Your estimates will be slightly different.)

4.4 How much does a spherical tank weigh if it must contain 285 lb of hydrazine with a 3% ullage? The maximum system temperature is 110°F; the minimum temperature is 40°F. The maximum stress in the walls is 110 ksi, and the maximum working pressure is 475 psia.

4.5 Design a hydrazine tank to meet the following requirements:
  1) Propellant load = 350 lb
  2) Maximum system temperature = 110°F
  3) Minimum system temperature = 40°F
  4) Blowdown ratio = 4.5
  5) Maximum operating pressure = 490 psia
  6) Propellant control device = 0.070-in.-thick elastomeric diaphragm.
  7) Allowable stress in titanium = 110,000 psi
What is the required volume of the tank and the estimated tank assembly weight? What is the weight of the helium loaded, assuming no absorption?

4.6 Design a titanium barrel tank, given:
  1) Allowable stress = 125 ksi
  2) Internal volume = 25. 3299 ft³
  3) Maximum tank pressure = 890 psia
  4) Inside diameter = 35.00 in.
(This is the Viking Orbiter tank; its actual length was 57.0 in., and its actual weight was 111.4 lb. Your estimate will be slightly different.)

4.7 A spacecraft blowdown, monopropellant hydrazine propulsion system has the following characteristics operating in a vacuum at a tank pressure of 410 psia:
  Thrust = 313 lb
  Chamber pressure = 345 psia
  Specific impulse = 233 s
What will thrust, chamber pressure, and specific impulse be when tank pressure drops to 310 psia?

4.8 Design a monopropellant hydrazine system for the following spacecraft:
Spacecraft dry weight = 1200 lb
I_x = 1500 slug-ft^2, I_y = 3000 slug ft^2, I_z = 8000 slug-ft^2
Maximum moment arm for thrusters = 6.5 ft (shroud internal radius)
The system must meet the following requirements:
Translational velocity change = 900 m/s (steady-state burns)
Reaction wheel unloading impulse = 9800 lb-s (steady-state burns)
Attitude-control maneuvers = 7000 lb-s (pulsing)
Steady-state I_{sp} = 230 s
Pulsing I_{sp} = 120 s
Minimum pulse width = 20 ms
Maximum time available for pitch, yaw, or roll maneuvers = 90 deg in 5 min
Blowdown pressurization; blowdown ratio = 4.5
Nitrogen pressurant
Maximum tank pressure = 650 psia
Diaphragm propellant control
Maximum stress in titanium = 100,000 psi
No single nonstructural malfunctions can cause loss of the mission.
Prepare a propellant inventory. Select the number of tanks. Select the thrust levels.
Size the tank(s). Size the diaphragm(s). Prepare a schematic; include instrumentation and first-pulse provisions. Prepare a propulsion weight statement. Prepare a failure modes and effects analysis.

5
Bipropellant Systems

Bipropellant systems offer the most performance (I_{sp} as high as 450 s) and the most versatility (pulsing, restart, variable thrust). They also offer the most failure modes and the highest price tags.

5.1 Bipropellant Rocket Engines

The major parts of a pressure-fed engine are the injector, the nozzle, and the cooling system and thrust chamber valves, as shown in Fig. 5.1. (Launch vehicle systems are pump-fed and considerably more complicated.) Oxidizer and fuel are fed as liquids through the injector at the head end of the chamber. Rapid combustion takes place as the liquid streams mix; the resultant gas flows through the converging-diverging nozzle.

Injector

The function of the injector is to introduce the oxidizer and fuel into the combustion chamber in such a way as to promote stable, efficient combustion without overheating the injector face or the chamber walls. The injector design is the single most important contributor to engine performance. It also determines to a large measure whether the engine will be stable. The most sensitive area in a bipropellant engine is the combustion zone of the thrust chamber just downstream of the injector. Propellants enter this high-pressure zone as room temperature liquids with velocities less than 200 fps and leave as gases at velocities greater than 2000 fps and temperatures as high as 6000°F. Energy release rates may be as great as 5000 hp/in.^{3} Mixing, ignition, and combustion occur in milliseconds. Propellants must be introduced through the injector plate at high velocity so that the fuel and oxidizer are shattered into droplets small enough to complete combustion upstream of the throat. Continuous and rapid combustion must occur near the injector face so that pockets of mixed but unreacted propellants do not form and explode. The thrust developed by a single pair of injector orifices can be as high as 500 lb. Figure 5.2 shows four of the myriad injector configurations in use.

Injector design, in spite of its importance, is empirical. The German team under Dornberger and von Braun had great difficulty in developing their first injector. After numerous mishaps, explosions, and instabilities, they got a nonimpinging showerhead design to work. For the larger V-2 engine, they decided simply to scale up their smaller, successful showerhead. The injector refused to work in a larger size and the old difficulties returned. After several attempts, they gave up on the scale-up model and used multiple small successful injectors assembled together. More than 4000 V-2s flew with this odd design.

The impinging jets like the doublet pattern work well with hypergolic propellants, which require rapid mixing and liquid phase combustion for high

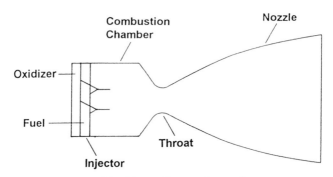

Fig. 5.1   Bipropellant rocket engine.

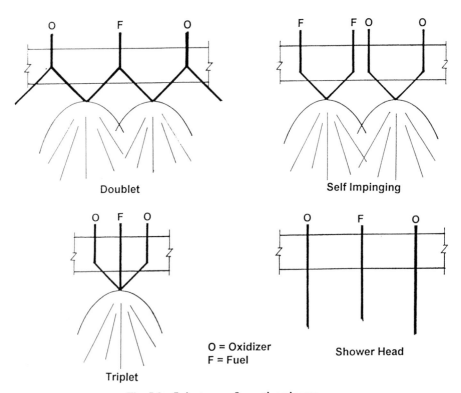

Fig. 5.2   Injector configurations in use.

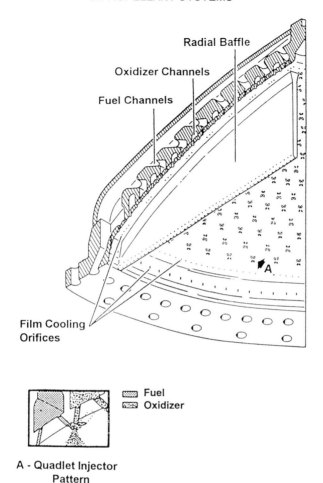

Radial Baffle

Oxidizer Channels

Fuel Channels

Film Cooling
Orifices

Fuel
Oxidizer

A - Quadlet Injector
Pattern

**Fig. 5.3   Titan IV, Stage II, injector head design. (Courtesy Aerojet Tech Systems
Company.)**

performance. The self-impinging design is used with hypergolic and nonhy-
pergolic propellant combinations. Impinging patterns are difficult to manufacture
because of the close tolerances required of the orifices. The showerhead design
works with cryogenics and is relatively simple to manufacture. The Titan IV, Stage
II, injector, shown in Fig. 5.3, uses a double doublet, or quadlet, pattern for the
hypergolic propellants nitrogen tetroxide and a mixture of hydrazine and UDMH.
In a quadlet, two streams of fuel and two streams of oxidizer are all directed at the
same point.

   The injector face contains 17 concentric rings that serve as the propellant
channels. The channels are alternately fuel and oxidizer. The outer ring is a fuel
ring and is a nonimpinging showerhead that directs film cooling fuel down the
chamber wall.

Cooling

The adiabatic flame temperature for nitrogen tetroxide and monomethylhydrazine is 5685°F; the heat transfer rates at the throat can be as high as 20 Btu/s-in.^{2} during operation. Clearly, throat and nozzle cooling is an important design problem. Heat transfer at the throat may be estimated from[31]

$$\frac{hd}{k} = 0.023 \left(\frac{V d\rho}{\mu} \right)^{0.8} \left(\frac{c_p \mu}{k} \right)^{0.4} \tag{5.1}$$

where

h = convection coefficient, Btu/s-ft^2-°F
d = diameter of the section, ft
V = velocity, fps
ρ = density, lb/ft-s
μ = viscosity, lb/ft-s
C_p = specific heat, Btu/lb-°F
k = thermal conductivity, Btu/ft-s-°F

Gas properties are a function of temperature and must be evaluated at the local temperature. The cooling methods used are:
1) Film cooling.
2) Regenerative cooling.
3) Radiation cooling.
4) Ablative cooling.
5) Beryllium cooling.

Film cooling is used in all engines as a supplement to the primary cooling method. A cylindrical film of fuel is injected into the chamber such that it washes down the wall. The flow is designed such that some liquid will reach the throat. Cooling comes from the heat of vaporization of the fuel. Combustion near the wall is fuel-rich and inefficient; therefore, specific impulse is reduced, and the amount of fuel used in film cooling is minimized in engine designs.

Regenerative cooling is used on all large launch vehicle engines. As shown in Fig. 5.4, the nozzle is constructed as a bundle of shaped cooling tubes brazed together and radially reinforced. The fuel is routed through the tubes from the exit plane up to the injector. The energy collected from the gas stream in cooling is returned to the combustion process by the hot fuel as it enters the chamber; hence, the process is regenerative. Some engines, notably the Titan engines, use double-pass regenerative cooling, where fuel is routed down from the injector head, collected in a manifold, and routed back. Downflow and upflow are in adjacent tubes. The area of regenerative cooling tubes is varied along the length to produce high-velocity flow in the hottest region, the throat.

In *ablative cooling*, the chamber wall is constructed of a material that will be evaporated and eroded away during the firing. Typically in such an engine, the throat will be a machined carbon block, and the chamber and diverging section will be a fiberglass epoxy material. The material thickness is sized so that the expected burn time will remove an allowable amount of material.

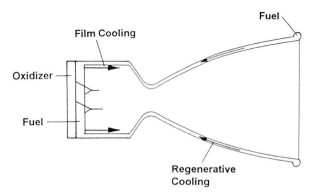

Fig. 5.4   Regenerative coolant flow.

For *radiation cooling*, the chamber is made of a refractory material, and radiative heat rejection to space plus enhanced film cooling keep the chamber walls within acceptable limits.

Beryllium cooling takes advantage of the remarkable heat-transfer properties of beryllium to transfer heat from the throat to the injector area, where film cooling can pick it up. The Viking Orbiter used a beryllium-cooled engine.

5.2   Bipropellants

Dr. Goddard, in his early work with rockets, chose liquid oxygen and kerosene for their availability and performance. When the Germans started their work with rockets, they chose oxygen but elected to use alcohol as the fuel because of the scarcity of petroleum. After World War II, when the work returned to the United States, kerosene was again the fuel of choice, renamed RP-1 (rocket propellant 1). Liquid oxygen was not a good choice for strategic weapons because of the difficulty storing it under field conditions. As a result, "storable propellants" were developed by the Air Force. Nitric acid was used as an oxidizer in very early work; however, in the early 1960s the Air Force selected less corrosive nitrogen tetroxide as the oxidizer. The fuels most seriously studied were hydrazine and unsymmetrical dimethylhydrazine (UDMH). Hydrazine was less stable but it delivered higher I_{sp}. The final selection for the Air Force launch vehicle program was a 50/50 mixture of hydrazine and UDMH. This combination, named Aerozine 50, was first used in Titan II and is the Titan propellant combination to this day. In later developments, the storable fuel of choice became monomethylhydrazine, or MMH, which has the properties of Aerozine 50 without the mixing problems.

Meanwhile, NASA, with its need for heavy launch vehicles, developed the Saturn series (Saturn 1, 1B, 5-Apollo, 5-Skylab) with oxygen/hydrogen. The Shuttle, in turn, was developed with this combination. Research continued with the more energetic fluorine oxidizer and derivatives like chlorine trifluoride, which were eventually dropped because of difficulty containing fluorine in metal tanks. The properties of some propellants are shown in Table 5.1.

Fluorine, oxygen and hydrogen are called *cryogenics* because of the extremely low temperatures at which they liquefy. There is a body of engineering techniques

Table 5.1 Properties of propellants

Propellant	Symbol	Molecular weight	Freezing point	Boiling point	Density @68°F	Vapor pressure, psia	°F
Chlorine trifluoride	ClF$_3$	92.46	−105.4	53.15	1.825	20.8	110
Fluorine	F$_2$	38	−363	−307	1.51[a]	5.0	−322
Hydrazine	N$_2$H$_4$	32.05	34.8	235.9	1.008	0.2	68
Hydrogen	H$_2$	2.02	−435	−423	0.071[a]	1.02	−435
MMH	CH$_3$N$_2$H$_3$	46.08	−62.1	188.2	0.8765	0.70	68
Nitric acid	HNO$_3$	63.02	−42.9	185.5	1.513	0.93	68
Nitrogen Tetroxide	N$_2$O$_4$ ——	92.02 ——	11.8 ——	70.1 ——	1.447 ——	13.92 ——	68 ——
Oxygen	O$_2$	32	−361.8	−297.6	1.14[a]	7.35	−308
RP-1	CH$_{1.9-2.0}$	175	−48	360 to 500	0.806	0.02	68
UDMH	(CH$_3$)$_2$N$_2$H$_2$	60.10	−71	146	0.793	2.38	68

[a]At normal boiling point.

called *cryogenic engineering* associated with the use of cryogenics. As one example, note that a liquid hydrogen tank must be insulated in such a way that air cannot penetrate the insulation. If air approaches liquid hydrogen temperatures, it will liquefy with resulting very high heat transfer to the hydrogen. As another example, if a cryogenic is allowed to warm in a long vertical feed line, it will form sudden large bubbles as it reaches the boiling point. These bubbles will coalesce and rise through the liquid with destructive force. This series of events is called *geysering*. One cure for geysering is to inject gaseous helium into the cryogenic, the cryogenic rapidly evaporates into the helium bubble seeking to reach a saturated vapor state; the result is rapid cooling of the cryogenic. Studies of the use of cryogenics in spacecraft have been conducted with the hope of using the cold temperatures achievable in space for cryogenic storage; however, cryogenics have never been used for anything other than launch vehicles; the Space Shuttle is considered a launch vehicle in this context.

Liquid Oxygen

Liquid oxygen (LOX) is light blue and has no odor; its boiling point is −297.6°F; it is stored and used at or near its boiling point. It does not burn but will support combustion vigorously. The liquid is stable; however, mixtures with fuel are shock-sensitive. It is not hypergolic; engines using this oxidizer require an igniter.

Liquid Hydrogen

Liquid hydrogen (LH$_2$) is transparent and odorless; its boiling point is −425°F. It is nontoxic but is extremely flammable; its flammable limits are 4% to 75% in air. Rocket exhaust, when burned with LOX, is colorless and difficult to see. LOX/hydrogen is the propellant combination used by the Shuttle's main engines.

The properties of hydrogen make it an excellent coolant and utility working fluid for engines.

Nitrogen Tetroxide

Nitrogen tetroxide (N_2O_4) is an equilibrium solution of nitrogen dioxide, NO_2, and nitrogen tetroxide, N_2O_4. The equilibrium state varies with temperature. It is a relative of nitric acid, which it smells like. It is reddish brown and very toxic. It is hypergolic (ignites spontaneously on contact) with hydrazine, Aerozine 50, and MMH; therefore, igniters are not required. This property makes pulsing performance practical with storable propellants. It is compatible with stainless steel, aluminum, and Teflon but incompatible with virtually all elastomers.

Monomethylhydrazine

Monomethylhydrazine (MMH) is a clear, water-white toxic liquid. It has a sharp decaying fish smell typical of amines. It is not sensitive to impact or friction; it is more stable than hydrazine when heated and will decompose with catalytic oxidation. Nitrogen tetroxide/MMH is the dominant spacecraft propellant combination for spacecraft propulsion.

5.3 Bipropellant Fluid Systems

Discussions in Chapter 4 of pressurization and propellant systems are equally applicable to bipropellant systems. Therefore, only the peculiarities of bipropellant systems will be discussed here.

Propellant Control

For spacecraft propulsion systems, nitrogen tetroxide and monomethyl hydrazine are the propellants of choice. The mixture ratio for maximum performance is about 1.6; however, the equal volume mixture ratio is 1.50. The equal volume mixture ratio is frequently used (Mariner 9, Viking Orbiter, Minuteman, Transtage) because a single tank design is adequate for both propellants. Titanium, aluminum, and stainless steel are compatible with both propellants and are the tank materials used; titanium is the lightest and most common.

Because nitrogen tetroxide is not compatible with elastomers, the propellant control devices are limited to metals and Teflon. Bellows, capillaries, and Teflon bladders and traps have been used.

The Minuteman postboost propulsion system used a stainless steel bellows assembly, shown in Fig. 5.5, to control propellant position and to accommodate long-term propellant storage. Minuteman, like most other projects, decided to make both oxidizer and fuel tanks essentially identical.

For the Space Shuttle reaction-control system, titanium tanks with complex stainless steel capillary systems, see Fig. 4.14, were chosen. For Mariner 9 and Transtage, the choice was Teflon bladders. In each case, the propellant control device choice was driven by the oxidizer but used in both systems. Propellant traps have had limited use. The Transtage used a propellant trap to provide a start slug for the main engine.

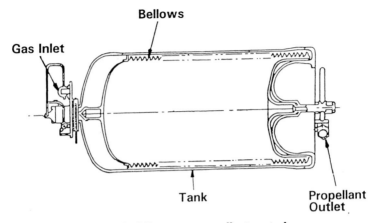

Fig. 5.5   Minuteman propellant control.

Pressurization

Blowdown pressurization has not been used with bipropellants because of 1) the difficulty in keeping both tanks at the same pressure and 2) difficulties with bipropellant engines with varying inlet pressure. Either helium or nitrogen systems are used although helium is the more frequent choice. A typical, simplified bipropellant pressurization system is shown in Fig. 5.6.

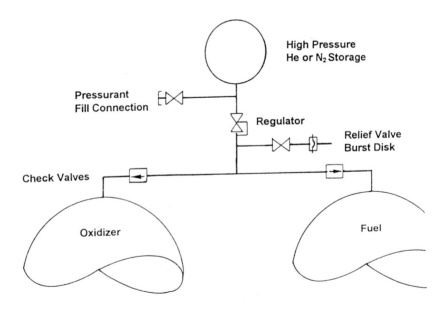

Fig. 5.6   Simplified bipropellant pressurization system.

Vapor mixing. Careful consideration should be given to propellant vapor mixing between the check valves after permeation of the check valve seals. Regulator leakage, caused by an obstruction on the valve seat, occurred as Viking 1 neared Mars. The spacecraft orbit insertion was successfully accomplished in spite of the leak, as was the remainder of the multiyear mission. The leakage was exhaustively investigated. The probable cause was propellant vapor mixing between the check valves. The same phenomenon was the most probable cause of the recent total failure of the Mars Observer. Positive mixing prevention is particularly important on long-duration missions.

Unusable Propellant

The unusable propellant is greater with bipropellant systems than monopropellant systems. The difference between loaded mixture ratio and burned mixture ratio results in a residual of one of the commodities. This residual is called *outage*. The unusable propellant is about 4% of the total load by weight.

Pulsing Performance

Sutton[7] states that an I_{sp} of 50% of theoretical can be expected with a bipropellant engine producing 0.01-s pulses (as compared to steady state of 92%). With 0.1-s pulses, 75% to 85% can be excepted. Reference [15] gives a pulsing I_{sp} for bipropellant engines of 170 s. Transtage motor testing with 0.02-s pulses indicates a delivered I_{sp} of 128 s or 45% of steady state, for a 25-lb engine and a delivered I_{sp} of 191 s, or 65% of steady state, for a 45-lb engine. Transtage engine firings produced an impulse repeatability of $\pm10\%$ pulse to pulse and $\pm30\%$ engine to engine.

5.4 Launch Vehicle Systems

Launch vehicle propulsion systems rest on the same body of theory as spacecraft systems but are quite different in implementation. For launch vehicles:

1) *Ambient pressure* is an important design consideration. In a spacecraft system, larger area ratio is always better. In a launch vehicle system, engines are designed for optimum area ratio at the average back pressure expected during the mission. A typical Stage I area ratio is about 10; a typical Stage II area ratio is about 20 compared to about 50 for vacuum engines. (Upper stage or Stage III engines are essentially vacuum engines.)

2) *High thrust level* is important, especially in Stage I. At liftoff, the Stage I thrust level must exceed the weight of the entire vehicle stack, including payload. If thrust does not exceed weight at ignition, the vehicle will rest patiently on the launch stand burning propellant until weight is reduced below thrust. It is desirable for the vehicle thrust-to-weight ratio at ignition to exceed 1.2; a net acceleration of 0.2 g will move the vehicle away from the launch tower with sufficient dispatch to make collision unlikely. Thrust to weight less than 1 is acceptable in upper stages. The importance of Stage I thrust led to the popularity of solid rocket first stages, which are characterized by high thrust and short burn time. Space Shuttle, Titan III, Titan IV, and Delta use strap on solids.

3) *High chamber pressure* is important. Recall that specific impulse is a function of chamber pressure when an ambient pressure is present and that high chamber pressure produces the most thrust for a given engine size. First-stage bipropellant engines typically have chamber pressures in the 500–1000-psi range.

Pump-Fed Systems

The net effect of high chamber pressure and the large tanks involved in launch vehicle propulsion dictate pump-fed engines for lower stages as opposed to pressure-fed spacecraft engines. In a pump-fed design, propellant is fed to the injector by a set of turbine-driven pumps. A typical pump-fed engine is shown in Fig. 5.7. Propellants are fed to the pump inlets at a pressure adequate to prevent

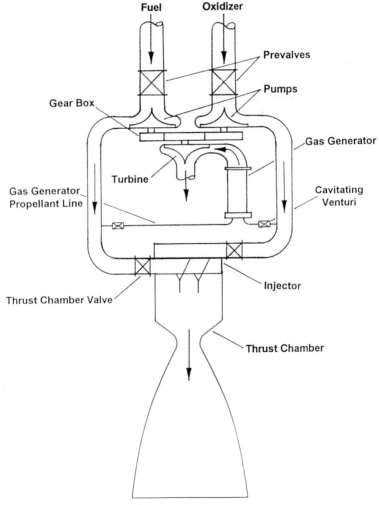

Fig. 5.7   Simplified pump-fed engine.

cavitation in the pumps. Turbine-driven rotary pumps increase propellant pressure to well above chamber pressure and deliver the required propellant flow rates to the engine injector. A small quantity of each propellant is bled from the pump outlet lines through flow-control devices (usually cavitating venturis) to a gas generator. Propellants are burned in the gas generator at an extremely fuel-rich mixture such the output exhaust gas is at an acceptable temperature, typically 1700° F. The gas generator exhaust energy is used to drive a turbine which, in turn, drives the oxidizer and fuel pumps through a gear box.

Start transient. The system shown in Fig. 5.7 is often called a *bootstrap system* because it will operate in steady state but will not start without assistance. To start the system, a solid propellant start cartridge is used to spin up the turbines that bring outlet pressures up to the point at which the gas generator can be lit. After the gas generator is lit, pump outlet pressure rapidly increases to a preset point at which the thrust chamber valves are opened in sequence to allow propellant flow to the injector. As the propellant flow starts through the injector, ignition occurs with hypergolic propellants, or an igniter is lit for cryogenic propellants. In neither case is propellant allowed to accumulate in the chamber. After ignition, flow rates and pressures continue to rise until the steady-state set point is reached.

Engine set point. The steady-state operating point of an engine can be controlled in two major ways:

Calibration. The main propellant flow rates and the gas generator propellant flow rates can be controlled by orificing the flow paths. With calibration firings, it can be ascertained that steady-state conditions are within specified limits under standard conditions. The individual engine performance under nonstandard in-flight conditions can then be predicted analytically.

Computer Control. The engine parameters can be actively controlled by a flight computer.

Calibration is by far the simplest method; however, computer control provides more versatility. The Titan engines are calibrated; the Space Shuttle main engines are computer-controlled.

Launch Vehicle Stage

The Titan IV, Stage I, a typical launch vehicle stage, consists of a rocket engine, an oxidizer tank, a fuel tank, pressurization equipment, and inner-stage structure; see Fig. 5.8. The rocket engine produces 548,000 lb of vacuum thrust. Each of the twin chambers is served by a turbine pump assembly and gas generator. The fuel and oxidizer tanks are welded aluminum with tapered chem-milled skins. The oxidizer load is 227,000 lb and the fuel load 119,000 lb. The stage burn time is 188 s; the stage empty weight is 18,700 lb. The stage mass ratio (weight of propellant divided by total weight) is 95%, about the same mass ratio as a loaf of bread.

Oxidizer is fed from the upper tank through a feed line that passes through a conduit in the fuel tank and branches to two individual engine lines. The lower tank, the fuel tank, is directly above the engine; fuel is fed to both engines through individual outlets. Both tank outlets are carefully designed so that propellant does not rotate and drop out as water in a kitchen sink does. It is important to consume the fuel in the feed lines without ingesting gas into the engine. Shutdown occurs

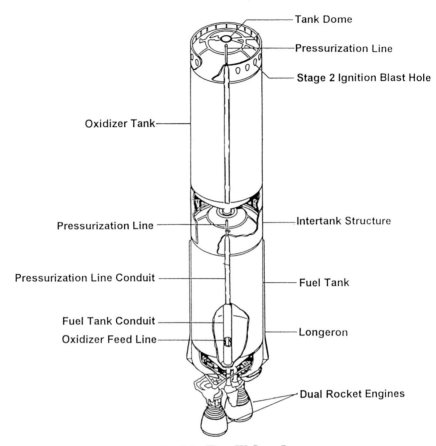

Tank Dome

Pressurization Line

Stage 2 Ignition Blast Hole

Oxidizer Tank

Intertank Structure

Pressurization Line

Pressurization Line Conduit

Fuel Tank

Fuel Tank Conduit

Oxidizer Feed Line

Longeron

Dual Rocket Engines

Fig. 5.8   Titan IV, Stage I.

on Titan stages when one of the propellants is depleted. Outage is the weight of propellant remaining in a tank when depletion occurs in the other tank.

Net positive suction head.   The purpose of the pressurization system in a launch vehicle is to assure that the suction pressure at each of the pumps is high enough to prevent cavitation in the pump. Cavitation occurs when the local pressure reaches the vapor pressure at any location. When cavitation occurs in a pump, catastrophic failure of the engine is likely because of the sudden drop in flow rate and discharge pressure or because of structural damage to the pump. Cavitation is prevented by supplying adequate net positive suction head at the pump inlet. Net positive suction head (NPSH) is defined as the pump suction pressure in excess of the propellant vapor pressure; suction pressure and vapor pressure are expressed in feet of head. Equation (5.2) shows how NPSH can be related to tank pressure.

$$\text{NPSH} = H_p + H_L - H_f - H_{\text{vp}} \qquad (5.2)$$

where

H_p = absolute tank pressure expressed as head, ft (absolute tank pressure
= gauge tank pressure + atmospheric pressure)

H_L = head due to length of liquid column over the pump; the vertical
dimension from the propellant surface to the pump centerline times
the current actual acceleration in g, ft

H_f = friction loss from the propellant surface to the pump inlet, ft

H_{vp} = vapor pressure of the propellant taken at the temperature of the
propellant entering the pump, ft

Note that the propellant head H_L is a function of burn time.
The required tank pressure is related to the required NPSH as follows:

$$P_t = P_v + \frac{\rho}{144}(\text{NPSH} - L_L a + H_f) \qquad (5.3)$$

where

P_t = tank pressure, psia
P_v = propellant vapor pressure at the pump inlet temperature, psia
L_L = length of liquid column from the propellant surface to the pump inlet, ft
a = instantaneous vehicle acceleration, g
ρ = propellant density, lb/ft^3

Example 5.1:

What tank pressure is required for a nitrogen tetroxide tank under the following
conditions:
Minimum NPSH = 44 ft
Pump inlet temperature = 80°F
Vertical height of liquid column above the pump inlet = 15 ft
Vertical acceleration = 1.8 g
Head loss due to friction = 3.5 ft

From Appendix B, the vapor pressure of nitrogen tetroxide at 80°F is 19 psia, and
the density of nitrogen tetroxide at 80°F is 89.242 lb/ft^3. From Eq. (3.3),

$$P_t = 19 + \frac{89.242}{144}[44 - (15)(1.8) + 3.5] = 39.5 \text{ pisa}$$

5.5 Dual-Mode Systems

Many missions require a high-impulse translation burn as well as pulse-mode
attitude-control operation. For example, a planetary orbiter requires a high-impulse
single burn for orbit insertion and pulse mode for attitude control. Magellan used
a solid motor for orbit insertion and a monopropellant system for pulse mode and
small translation burns. Galileo used a bipropellant system for high-impulse burns
and for pulse-mode operation. Viking Orbiter used an MMH/N$_2$O$_4$ bipropellant
system for orbit insertion and trajectory correction maneuvers, and a cold-gas

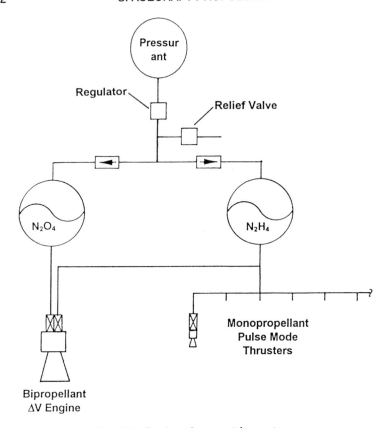

Fig. 5.9   Dual-mode propulsion system.

system for attitude control. There is an emerging technology that promises a better solution by using hydrazine as bipropellant fuel, replacing monomethylhydrazine, and also as a monopropellant for pulsing. This is called the *dual-mode system*. The system is shown schematically in Fig. 5.9.

The advantages of the dual-mode system are: 1) the ability to use the hydrazine as a monopropellant in attitude-control thrusters and as the fuel in bipropellant main engines; and, with resulting system simplification, 2) an increase in I_{sp} of several seconds. The disadvantage is that the system has minimal flight history. Hydrazine is less stable than MMH, and hydrazine bipropellant engines would be somewhat more likely to have instabilities. Instability can be prevented, however, with adequate engine development. The dual-mode system has been chosen for the Mars Global Surveyor spacecraft scheduled for launch in 1996.

5.6   Flight Bipropellant Systems

Five spacecraft bipropellant systems with a successful flight history are described in this chapter; see Table 5.2. The systems are arranged in the order of first flight to show the progression of technology.

Table 5.2 Spacecraft bipropellant systems

	Transtage RCS	Viking Orbiter	Shuttle RCS	Galileo	Intelsat VI	Mars Global Surveyor
First launch	1964	1975	1981	1989	1989	1996
No. thrusters	8	1 (ACS by cold gas)	44	13	8	13
Thrust, lb	25, 45	300	25, 870	2.25, 90	5, 110	1, 134
Engine cooling	Ablative	Beryllium	Radiation cooled and insulated	Radiation	Radiation	Radiation
Fuel	50/50 mix of hydrazine and UDMH	MMH	MMH	MMH	MMH	Hydrazine
Oxidizer	Nitrogen tetroxide	Nitrogen tetroxide	Nitrogen tetroxide	Nitrogen tetroxide	Nitrogen tetroxide	Nitrogen tetroxide
Mixture ratio	1.60	1.50		1.6	1.6	
Propellant control	Teflon diaphragms	Capillary vane devices	Capillary screens	Centrifugal 10(rpm)	Centrifugal	Capillary vane
Propellant tanks	Titanium equal volume spherical	Titanium equal volume barrel	Titanium equal volume, spherical	Four equal volume, titanium, spherical	Eight equal volume, titanium, spherical	Three equal volume, titanium, barrel
Pressurization	Regulated nitrogen	Regulated helium		Regulated helium	Regulated helium	Regulated helium
Vapor mixing prevention	Single, soft seat check valves	Series soft seat check valves		Single soft seat check valves, low leak design	Single check valves	Pyro-valves
Dry mass, lb	55	442		2040		139
Propellants, lb	120	3137			5100 to 5990	836
		32		33	34	
Primary reference features	Early design	Beryllium cooling	Large size, multiuse	Spinner, flushing burns	Spinner, redundant half-systems	Dual-mode operation

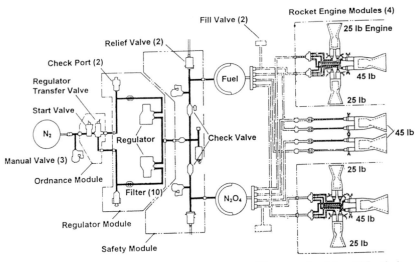

Fig. 5.10   Transtage reaction control system. (Courtesy Lockheed Martin.)

Transtage Reaction-Control System

The Transtage system was one of the first bipropellant reaction-control systems ever flown. The system provided propellant settling for main engine starts and attitude control during unpowered coasts. The system, shown schematically in Fig. 5.10, used nitrogen tetroxide and a 50/50 mixture of hydrazine and UDMH. The fuel is a mechanical mixture with properties and performance similar to MMH. It is the fuel used on all of the Titan launch vehicles. The system consists of four 45-lb pitch and yaw thrusters and four 25-lb roll-control thrusters. The engines use ablative cooling with integral mounting structure at the throat and head end. The performance of the engines is summarized in Table 5.3. The nominal propellant load was 74 lb of oxidizer and 46 lb of fuel. Titanium tanks and

Table 5.3   Transtage-reaction-control thrusters

	Roll motor	Pitch motor
Thrust, lb	25	45
Chamber pressure, psia	155	140
Min. steady state I_{sp}, s	285	285
Nominal steady-state I_{sp}, s	295.6	296.7
Min. 20-ms pulsing I_{sp}, s	128	191
Min. impulse bit, lb-s	0.375	0.675
Repeatability bit-bit, ± lb-s	0.03	0.054
Repeatability eng-eng, ± lb-s	0.083	0.217
Mixture ratio	1.56	1.56

Teflon diaphragms were used for each propellant. The regulated pressurization system used nitrogen gas in a single titanium sphere. Pressurant and propellant vapors were kept separate by single, Teflon-seat check valves between the propellant tanks. Separate pressure-relief valves are provided for each propellant tank.

The bipropellant system was replaced with a monopropellant hydrazine system in the late 1960s, when the spontaneous catalyst became available.

Viking Orbiter

The Viking Orbiter was the spacecraft that carried the Viking Lander to Mars, installed it in Mars orbit, took high-resolution pictures of potential landing sites, and released the Lander, on cue, for entry and landing. After releasing the Lander, the Orbiter served as a relay radio link to Earth and conducted an independent mission of its own, videomapping the Mars surface at higher resolution than achieved before or since.

The Viking Orbiter propulsion system was a fixed-thrust, multistart, pressure-fed, storable bipropellant system designed to deliver 10^6 lb-s of impulse to the Viking spacecraft. Most of the impulse was required for Mars orbit insertion; however, it also provided course corrections near Earth and near Mars as well as periodic Mars adjustment burns. (Attitude control for the Orbiter was provided by a separate nitrogen cold-gas system.) The system schematic is shown in Fig. 5.11.

The propellants were nitrogen tetroxide and monomethylhydrazine. Regulated gaseous helium was used for pressurization. The system was powered by a 300-lb (vacuum) thrust beryllium-cooled engine. The engine was gimbaled; electromechanical actuators provided thrust vector control in pitch and yaw during engine firing. Roll control was provided by the small nitrogen cold-gas thrusters. The tanks were identical, an economy that results in operation at equal volume mixture ratio. The tanks contain capillary propellant management devices. These were the first vane-type devices to fly. The design and performance characteristics of the Viking Orbiter propulsion system are shown in Table 5.4.

The pyrovalve ladder, shown at the helium tank outlet (colloquially known as the "minefield") was used to open and reseal the helium source between groups of burns; its purpose was to prevent loss of helium from leakage during long coast periods. The system was welded and brazed to prevent fluid leakage. There were no mechanical joints.

Two systems were launched to Mars in 1975, each carrying a Viking Lander. Both systems performed the mission satisfactorily although, regulator leakage occurred on the first Orbiter as it neared Mars. It was necessary for the MOS team to take corrective action to avoid excessive tank pressure. The most probable cause of the regulator leakage is believed to be vapor mixing between the check valves, which produces a gummy residue in the same passage with the regulator. Regulator leakage decreased with use, and the remainder of the Viking 1 mission and all of the Viking 2 mission were flown nearly perfectly.[35] The Viking legacy requires that bipropellant propulsion system designs preclude propellant vapor mixing.

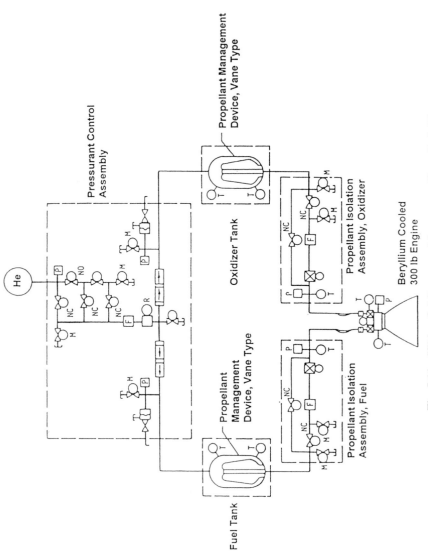

Fig. 5.11 Viking Orbiter propulsion system (from Ref. 32).

Table 5.4 Viking Orbiter bipropellant propulsion system (from Ref. 32)

Performance summary		Mass summary	
Parameter	Value	Item	Mass, lb
Vacuum thrust	300 ± 9 lb	Structure	43.0
Vacuum I_{sp}	291.4 s	Assembly hardware	26.9
Area ratio	60:1	Pressure-control assembly	28.1
Chamber pressure	115 ± 3.4 psia	Propellant tanks, two	222.8
Mixture ratio	1.50 ± 0.056	Propellant isolation assembly	24.8
Oxidizer flow rate, nom.	0.618 lb/s	Rocket engine assembly	18.6
Fuel flow rate, nom.	0.412 lb/s	System dry mass	442.7
Propellant load capacity	3137 lb	Pressurant	10.3
Usable propellant	3097 lb	Propellant trapped	40.4
Min. duration	0.4 s	System burnout mass[a]	493.4
Shutdown impulse, 3σ	2 to 12 lb/s		
Oxidizer	N_2O_4		
Fuel	MMH		
Pressurant	He		
Roll torque, max.	4.0 lb		

[a]Includes gimbal actuators, thermal control, cabling, unusable propellant.

The Viking Orbiter propulsion system was designed by Lockheed Martin under contract to the Jet Propulsion Laboratory.

Space Shuttle Reaction-Control System

The Shuttle reaction-control system consists of 44 bipropellant engines that maneuver the orbiter in space. There are 38 primary thrusters, each with 870 lb of thrust. There are six 25-lb vernier engines. The RCS thrusters are grouped in three modules, one in the nose, shown in Fig. 5.12, and one on each aft OMS pod. Each thruster group has its own set of capillary-controlled propellant tanks; the capillary devices are shown in Fig. 4.14. The propellants are nitrogen tetroxide and monomethylhydrazine. Propellants can be shared with the OMS propulsion system in the event of a failure. Figure 5.13 shows the insulated 25-pound engine. The nozzle is cut at an angle to fit the contour of the Shuttle skin. The vacuum I_{sp} is 260 s (min.) at a propellant inlet pressure of 246 psia; chamber pressure is 110 psia. The thruster can deliver pulse widths of 0.08 to 0.32 s, with a minimum offtime of 0.08 s, or it can be fired steady-state.

The Space Shuttle reaction-control system was designed by Rockwell International under contract to NASA.

Galileo

Galileo is the next step in the exploration of Jupiter following the flyby missions of Pioneers 10 and 11 and Voyagers 1 and 2. The Galileo mission is to find out

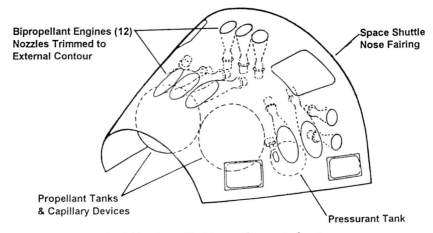

Fig. 5.12   Space Shuttle reaction-control system.

more about the chemical composition and physical state of the atmosphere and to study the satellites of Jupiter. The spacecraft is a dual-spin orbiter and an attached probe. Galileo was the second interplanetary spacecraft to be launched on the Shuttle following the 1989 Magellan flight.

The Galileo propulsion system is a highly redundant bipropellant MMH/nitrogen tetroxide system shown in Fig. 5.14. The system provides midcourse corrections, orbit adjustments, and Jupiter orbit insertion, which occurs after release of the probe. Maneuvering propulsion is accomplished with a single 400-N main engine; attitude control is provided with two clusters of six 10-N thrusters. The two clusters are on opposing booms. The 400-N engine can be fired for up to 70 min; the 10-N thrusters can be used in pulse mode or steady state. The minimum impulse bit is 0.09 N/s, with an ontime of 22 ms. The 10-N thrusters can back up the 400-N engine in the event of a failure. Three parallel supply branches feed propellant to the two redundant thruster branches and to the 400-N engine. The thruster

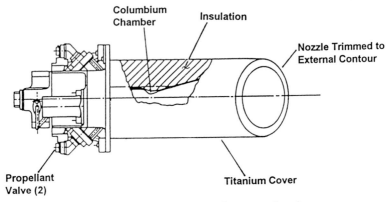

Fig. 5.13   Space Shuttle reaction-control engine.

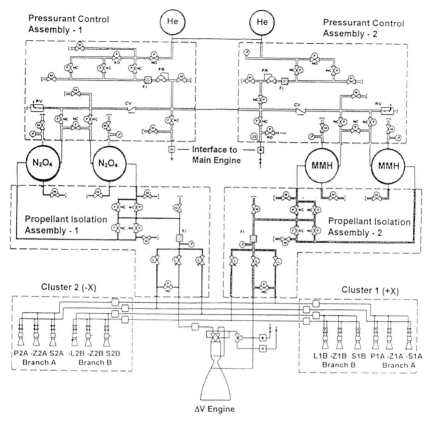

Fig. 5.14   Galileo propulsion system (from Ref. 33, p. 4).

arrangement and branch membership is shown in Fig. 5.15; the function of the thrusters is shown in Table 5.5.

Dual helium tanks provide helium for the regulated pressurization of the four propellant tanks. Special low-leakage check valves were designed to prevent propellant vapor mixing between the two check valves. The system incorporates periodic thruster firings (for example, at least every 23 days, all 10-N thrusters are fired for a minimum of 1.2 s) to flush the propellant that has been in contact with stainless steel parts. There is some evidence to indicate that these flushes are necessary to eliminate buildup of products from the interaction of nitrogen tetroxide and stainless steel.[36] Although the propellant system is primarily titanium, there are some stainless steel parts.

The Galileo propulsion system was designed by MBB and was provided to JPL/NASA by the then Federal Republic of Germany.

INTELSAT VI

The INTELSAT VI spacecraft are the latest in a long line of INTELSAT communication satellites, the first of which was launched in October 1989. The propulsion

SPACECRAFT PROPULSION

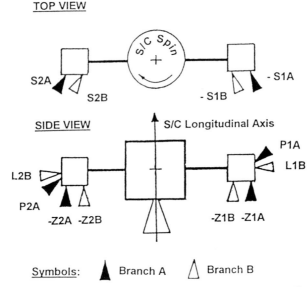

Fig. 5.15   Galileo thruster arrangement (from Ref. 33, p. 5).

system, shown in Fig. 5.16, supplies thrust for apogee injection (replacing a kick motor) and orbital adjustments.

The system consists of two functionally independent half-systems connected by latch valves on the liquid side and normally open squib valves on the pressurant side. Each half-system can perform all the propulsion functions required for spacecraft operation. Individual fuel and oxidizer tanks can be isolated on the pressurant side to prevent propellant migration. Two 110-lb liquid apogee motors (LAMs) provide the impulse for geosynchronous orbit insertion and spacecraft reorientation during the orbit transfer portion of the mission. The LAMs are located in the aft end of the spin-stabilized spacecraft and are used in steady-state

Table 5.5   Thruster functions (from Ref. 33, p. 5).

Maneuver	Thrusters used	
	Branch A[a]	Branch B
Attitude control		
Spin-up	S2A	S2B
Spin-down	−S1A	−S1B
Precession control	P1A, P2A	L1A, L2B
Turns	−Z1A, −Z2A	−Z1B, −Z2B
Trajectory control		
Longitudinal ΔV	−Z1A, −Z2A	−Z1B, −Z2B
Lateral ΔV	P1A, P2A	L1B, L2B

[a]Primary.

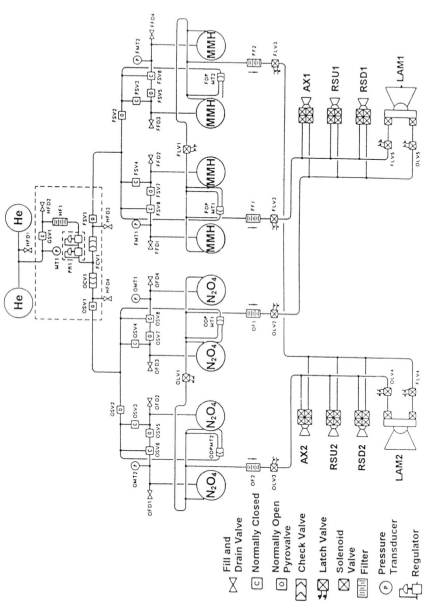

Fill and
Drain Valve

Normally Closed

Normally Open
Pyrovalve

Check Valve

Latch Valve

Solenoid
Valve

Filter

Pressure
Transducer

Regulator

Fig. 5.16 INTELSAT VI propulsion system (from Ref. 34, p. 4).

or pulse mode. Six 5-lb reaction-control thrusters are used for control functions. Two are used for axial impulse and are located at the aft end of the spacecraft near the LAMs. Four thrusters are located radially on the spun shelf and are used for spin-up and spin-down. The thrusters also provide impulse precession, orbit trim, nutation control, north-south stationkeeping and east-west station change, and deorbit control. The thrusters are separated into two redundant groups.

Helium pressurization is supplied through series-redundant regulators. Eight propellant tanks are symmetrically located around the center of mass. Both the oxidizer and fuel tanks are manifolded to maintain spacecraft balance. Centrifugal forces from spinning provide propellant position control; no propellant management device is necessary.

When the propellants and pressurant are launched, they are isolated from one another by latching valves and pyrovalves. Initial orientation and control is provided by the 5-lb thrusters in the blowdown mode. The pressurization system is opened for the apogee burn. After the apogee burn and trim maneuvers, the LAMs are isolated from the propellant supply by closing the isolation latch valves (FLV4 and 5 and OLV4 and 5) to preclude leakage failure. Sometime after LAM isolation, the pressurization system is sealed, and oxidizer and fuel systems are isolated to prevent propellant vapor mixing or propellant migration. After isolation, the system operates in blowdown mode for the remainder of the mission. Mixture ratio can be controlled during blowdown by operating FSV1 and OSV1.[38]

The Intelsat VI propulsion system was designed by Hughes Space and Communications Company under contract to International Telecommunication Satellite Organization (INTELSAT). Five Intelsat VI spacecraft were launched; each propulsion system performed satisfactorily.[34]

Mars Global Surveyor

The Mars Global Surveyor, scheduled to be launched in 1996, is a significant element of the NASA overall strategy for the exploration of Mars over the next decade. The propulsion system, shown in Fig. 5.17, provides Mars orbit insertion, attitude-control pulsing, and a number of orbit trim maneuvers. The system is one of the first to use the dual-mode concept that provides hydrazine fuel for bipropellant use during ΔV burns and for monopropellant pulsing.

The system uses three identical titanium barrel tanks (two fuel, one oxidizer) with capillary vane propellant positioning systems patterned after the Viking Orbiter. The tanks are sized to accommodate a 836-lb total propellant load. First-pulse problems are prevented by launching with propellant loaded to the thrust chamber valves.

Helium pressurant is stored in a seamless aluminum container with a graphite composite overwrap; the tank contains 3.0 lb of helium at 4200 psi. Vapor migration is prevented by a ladder of pyrovalves in the oxidizer pressurization line. These valves allow the oxidizer tank to be isolated between bipropellant burns. In addition, specially designed, low vapor leak, quad-redundant check valves are used in the pressurization system to reduce vapor migration during periods when the pyrovalves are open.

The Mars Global Surveyor spacecraft is being designed by Lockheed Martin under contract to the Jet Propulsion Laboratory.

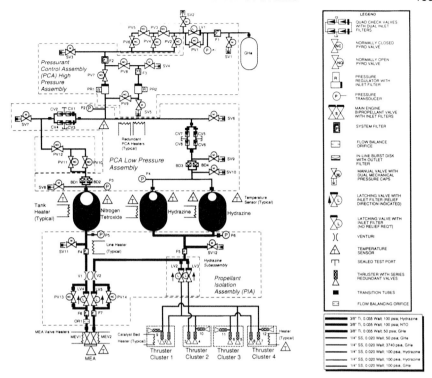

Fig. 5.17   Mars Global Surveyor dual-mode propulsion system. (Courtesy Lockheed Martin.)

Problems

5.1   The Titan IV, Stage I, is loaded with 227,000 lb of nitrogen tetroxide and 119,000 lb of aerozine 50. During the mission, the actual burned mixture ratio was 1.915. How much propellant, by weight, was left at the end of the burn? Which propellant caused the depletion shutdown?

5.2   The mixture ratio for an engine is 2.2. What is the volumetric mixture ratio if the propellants are MMH and nitrogen tetroxide?

5.3   Prepare a propellant inventory for a bipropellant system that will deliver a ΔV of 2.34 km/s to a spacecraft with a burnout weight of 988 lb (including unused propellant). The propellants are nitrogen tetroxide and MMH consumed at a mixture ratio of 1.55; the engines deliver an I_{sp} of 312 s. Use the following:

Propellant reserve = 5% of usable

Trapped = 3% of usable

Loading error = 0.5% of usable

Outage = 1.5% of usable

5.4   Design a bipropellant propulsion system. Assume:
Propellants are nitrogen tetroxide and MMH
Loaded propellant weight $= 1394$ kg
Nitrogen-regulated pressurization
Nitrogen stored at 3000 psia initially and 500 psia at burnout
Titanium propellant and pressurant tanks, allowable stress $= 96,000$ psi,
   titanium density 0.16 lb/in.^{3}
Propellant tank maximum pressure $= 200$ psig
Minimum system temperature 50°F
Equal volume mixture ratio
Ullage $= 3\%$ of tank volume initially
Determine:
Weight of propellant and pressurant tanks
Weight of pressurizing gas
Propulsion system dry weight if valves, engines, and plumbing weight 150 lb

5.5   Compare the weight of a solid motor and a bipropellant liquid system for
the Mars orbit insertion burn.
   a. Calculate the propellant weight for a solid motor system and an N_2O_4/MMH
bipropellant system for the following requirements:
- Mars orbit insertion $\Delta V = 2.7$ km/s
- Solid motor; $I_{sp} = 290$ s
- Bipropellant; $I_{sp} = 312$ s
- Spacecraft on orbit dry weight $= 1000$ kg

   b. Design a N_2O_4/MMH bipropellant engine. Assume that four 300-lb engines
are used, with an area ratio of 90 and a chamber pressure of 300 psi. Use the equal
volume mixture ratio.
   c. Prepare the propellant inventory.
   d. Size the propellant tanks. Assume:
3% ullage
Propellant tank pressure, max. $= 480$ psi
Teflon diaphragm propellant control devices, thickness 0.060 in.
All tanks titanium with an allowable stress of 100,000 psi
Size for 80°F propellant; size tanks using the oxidizer
   e. Size the pressurant tank. Assume:
Regulated pressurization
Helium pressurant
   Initial pressure, max. $= 5000$ psi
   Final pressure $= 550$ psi min.

   f. Prepare a complete weight statement for the bipropellant systems.
   g. Using the PRO software, design an equivalent PBAN/AP/AL solid rocket
motor for Mars orbit injection. Assume a chamber pressure of 500 psi and an area
ratio of 40. The burn time must be less than 100 s.

6
Solid Rocket Systems

In a solid motor, the oxidizer and fuel are stored in the combustion chamber as a mechanical mixture in solid form. When the propellants are ignited, they burn in place. Solid rocket systems are used extensively in situations in which 1) the total impulse is known accurately in advance and 2) restart is not required. Boosters for Titan IV and the Space Shuttle are good examples of such a situation. Spacecraft examples are the kick stage for a geosynchronous orbiter and the orbit insertion motor for a planetary orbiter. Solid motors are used in applications requiring impulses as small as 0.5 million lb-s (Aircraft JATO) and as large as 326 million lb-s (one Shuttle booster motor). Diameters range from 1 to 260 in.

The elements of a solid rocket motor, shown in Fig. 6.1 are discussed in subsequent sections of this chapter. In addition, the spacecraft equipment required to support a solid rocket motor is discussed.

6.1 Propellants

Solid rockets are over 700 years old, having been used by the Chinese in 1232. The basis of Chinese rockets was ordinary black powder, the gunpowder of muzzle-loaders. Smokeless powder, invented by Nobel in 1879, was not used as solid rocket fuel until Dr. Goddard tried it in 1918. He concluded that liquid propellants would be the best development path for space work, and his conclusion held true for 50 years. *Double-base* propellant, which is the interesting combination of nitrocellulose and nitroglycerin, was used for a time, and combinations of composite and double-based propellants have been used. Modern high-performance solid rockets were made possible in 1942 when the Jet Propulsion Laboratory introduced the first *composite propellant*. By 1960, composite propellants were the industry standard.

Composite propellants are composed of one of several organic binders, aluminum powder, and an oxidizer, which is nearly always ammonium perchlorate (NH_4ClO_4). The binders are rubberlike polymers that serve a dual purpose as fuel and to bind the ammonium perchlorate and aluminum powders into a solid capable of being shaped into a grain. Some binders are

CTPB = carboxy-terminated polybutadiene
HTPB = hydroxy-terminated polybutadiene, which is used in the Thiokol Star series of motors
PBAA = polybutadiene acrylic acid
PBAN = polybutadiene acrylonitrile, which is used in the Titan solid rocket motors
PU = polyurethane
PS = polysulfide
PVC = polyvinyl chloride

135

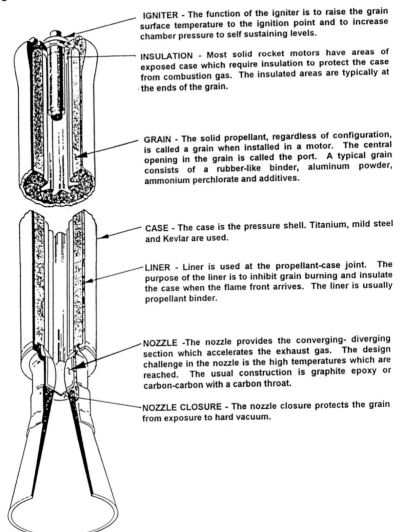

IGNITER - The function of the igniter is to raise the grain surface temperature to the ignition point and to increase chamber pressure to self sustaining levels.

INSULATION - Most solid rocket motors have areas of exposed case which require insulation to protect the case from combustion gas. The insulated areas are typically at the ends of the grain.

GRAIN - The solid propellant, regardless of configuration, is called a grain when installed in a motor. The central opening in the grain is called the port. A typical grain consists of a rubber-like binder, aluminum powder, ammonium perchlorate and additives.

CASE - The case is the pressure shell. Titanium, mild steel and Kevlar are used.

LINER - Liner is used at the propellant-case joint. The purpose of the liner is to inhibit grain burning and insulate the case when the flame front arrives. The liner is usually propellant binder.

NOZZLE -The nozzle provides the converging- diverging section which accelerates the exhaust gas. The design challenge in the nozzle is the high temperatures which are reached. The usual construction is graphite epoxy or carbon-carbon with a carbon throat.

NOZZLE CLOSURE - The nozzle closure protects the grain from exposure to hard vacuum.

Fig. 6.1   The Elements of a solid rocket motor.

Composites also contain small amounts of chemical additives to improve various physical properties, for example, to improve burn rate; promote smooth burning (flash depressor); improve casting characteristics; improve structural properties (plasticizer); and absorb moisture during storage (stabilizer).

A composite propellant is described by an acronym of the form: (binder)/AP (ammonium perchlorate)/AL (aluminum). For example, the Titan booster solid rocket propellant is PBAN/AP/AL. Since aluminum and ammonium perchlorate are essentially a given, the acronym is often shortened to the binder; the Titan booster composite propellant is commonly called simply PBAN.

In general, solid propellants produce specific impulse of 300 s or lower, somewhat lower than bipropellants. Solid propellant densities, however, are higher, resulting in smaller systems for a given impulse. As higher specific impulse is sought, less stable ingredients, such as boron and nitronium perchlorate, are necessary.

Hazard Classification

Propellants are divided into two explosive hazard designations by the DOD and DOT. In DOD Class 7 (or DOT Class A), catastrophic failure produces detonation. Some double-base and some double-base plus additive propellants are Class 7. In DOD Class 2 (or DOT Class B), catastrophic failure produces burning or explosion. Almost all common propellants are Class 2; some examples are CTPB/AP/AL, PBAN/AP/AL, HTPB/AP/AL, and PBAA/AP/AL.

The motor classification dictates the safety requirements for using and handling the motor. A Class 2 motor is obviously the least expensive motor to transport and handle.

Burning Rate

Control of the exhaust gas flow rate is achieved, not by precise metering of propellant flows as in a liquid system, but by precise control of the exposed grain surface area and the burning rate of the propellant mixture. The solid is transformed into combustion gases at the grain surface. The surface regresses normal to itself in parallel layers; the rate of regression is called the burning rate. The mass flow rate of hot gas leaving the motor is proportional to the product of area burning, the grain density, and the burning rate:

$$\dot{w} = A_b \rho_g r \qquad (6.1)$$

where

$\dot{w}$ = instantaneous hot gas flow rate through the nozzle, lb/s
A_b = instantaneous grain surface area at the flame front, in.²
ρ_g = density of the unburned propellant grain, lb/in.³
r = burning rate, the speed at which the flame front is progressing, in./s

The thrust-time curve of a solid rocket motor can be estimated by assuming that I_{sp}, grain density, and burning rate are constant and by calculating instantaneous values of flow rate and, hence, thrust:

$$F = \dot{w} I_{sp} \qquad (6.2)$$
$$F = A_b \rho_g r I_{sp} \qquad (6.3)$$

Between iterations of Eq. (6.3) burning rate can be used to calculate the next position of the burning surface from which the new area can be calculated.

Example 6.1: Thrust-Time Calculation

Consider a propellant grain 12 in. in diameter and 24 in. long with a cylindrical port. The grain density is 0.064 lb/in.³, the burning rate 0.55 in./s, and the vacuum

Flame Front Velocity =
Burning Rate, r, in/sec

Fig. 6.2   Burning rate measurement.

specific impulse 290 s. Calculate the vacuum thrust at the end of the third second of burning, when the port diameter is 6 in., and at the start of the fourth second of burning; assume that both ends of the grain are inhibited.

At the end of 3 s, the area burning is

$$A_b = \pi(6)(24) = 452 \text{ in}^2$$

and

$$F = A_b \rho_g r I_{sp}$$
$$F_3 = (452)(0.064)(0.55)(290) = 4614 \text{ lb}$$

By the start of the fourth second of burning, the diameter of the port will be $6 + (2)(0.55) = 7.1$ in. the port area and thrust will be

$$A_b = \pi(7.1)(24) = 535 \text{ in}^2$$
$$F_3 = (535)(0.064)(0.55)(290) = 5461 \text{ lb}$$

Burning rate can be determined by measuring flame-front velocity on a strand, as shown in Fig. 6.2.

Pressure effects.   Since grain density is essentially constant during a burn, gas flow rate is controlled by grain area and burning rate. Gas flow rate, thrust, and chamber pressure increase with an increase in burning area. Also, burning rate increases exponentially with pressure,

$$r = cP^n \qquad (6.4)$$

where

c = empirical constant influenced by grain temperature
n = burning rate pressure exponent

Table 6.1 shows the burning rates and pressure exponents for common propellant combinations. A zero pressure exponent would indicate, a propellant without burning rate sensitivity to pressure. A pressure exponent approaching 1 indicates high burning rate sensitivity to chamber pressure and potential for violent pressure rise in the chamber.

Chamber pressure.   By assuming that the change in mass of the gas in the chamber is negligible, it can be shown that chamber pressure and burning surface area are related as follows:[7]

$$P_c = \left(\frac{A_b}{A_i} \right)^{\frac{1}{1-n}} = K^{\frac{1}{1-n}} \qquad (6.5)$$

Table 6.1   Burning rates and pressure exponents

Propellant	Burning rate, in./s	Pressure exponent
CTPB/AP/AL	0.45	0.40
HTPB/AP/AL	0.28	0.30
PBAA/AP/AL	0.32	0.35
PBAN/AP/AL	0.55	0.33
PS/AP/AL	0.30	0.33
PVC/AP/AL	0.45	0.35

where

A_b = grain surface area burning
A_t = nozzle throat area
n = burning rate pressure exponent
K = ratio of burning area to throat area

Temperature effects.   Figure 6.3 shows the effect of grain initial temperature on chamber pressure and, hence, thrust. The sensitivity of chamber pressure to temperature is typically 0.12% to 0.50 % per °F.

Total impulse is essentially unchanged by initial grain temperature; however, thrust and loads on the spacecraft increase with temperature. Burning time decreases with temperature. Spacecraft motor temperature is carefully controlled for two reasons: 1) to limit variation in performance and 2) to minimize thermal stress in the grain and, hence, minimize the possibility of grain cracks. A typical

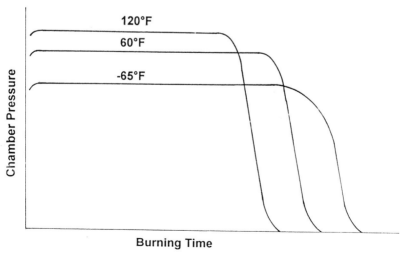

Fig. 6.3   Effect of grain temperature on chamber pressure.

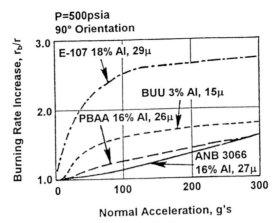

Fig. 6.4   Acceleration effect on burning rate (From Ref. 39, p. 50).

temperature-control range is 60 to 100°F; the control range may be tightened for a period just before the motor fires. Magellan used electrical heaters, thermostats, and insulation to control the motor temperature.

Acceleration effect.   Burning rate is increased by acceleration perpendicular to the burning surface, a phenomenon of particular interest in the design of spinning spacecraft.[39] Figure 6.4 shows that acceleration in the range of 30 *g*'s can double the burn rate. The enhancement of burning rate has been observed for acceleration vectors at any angle of 60 to 90 deg with the burning surface. The increase in burning rate is attributed to the presence of molten metal and metal oxide particles, which are retained against the grain surface by the radial acceleration, with an attendant increase in heat transfer at the grain surface. The effect is enhanced by increased aluminum concentration and particle size.[39,40]

Space Storage

Although there is some controversy and sparse data regarding the ability of a solid motor grain to withstand long periods of space exposure with an open port. The concern is that evaporation could affect the chemical composition of the propellant or bond line and thus impair performance. To address this concern, two options are available: 1) Conduct adequate testing to establish postvacuum performance or 2) design a sealed nozzle closure.

Most missions, kick stages, and upper stages cause only short-duration exposure. These missions have been flown for years with no closure. A planetary mission, like Magellan, is another story. Morton-Thiokol space aging data[29] for TPH-3135 shows low bond line and propellant risk for 10-month exposure. LDEF results[29] indicate that 5.5-year exposures may be acceptable. The Magellan mission itself shows (one data point) that 15-month space exposure without a nozzle plug is acceptable.

6.2   Grains

In a solid motor, the propellant tank and the combustion chamber are the same vessel. The viscous propellant mixture is cast and cured in a mold to achieve the desired shape and structural strength. After casting, the propellant is referred to as the *grain*. The most frequent method of casting is done in-place in the motor case using a mandrel to form the central port. When cast in this way, the grain is called case-bonded.

Grains may also be cast separately and loaded into the case at a later time; such grains are called *cartridge-loaded*. Large grains are cast in segments, the segments are then stacked to form a motor. The motors for the Shuttle and Titan IV are made in this way; Fig. 6.5 shows the segmented Shuttle motor. Segmented grains solve the difficult motor transport problem; however, they also require high-temperature case joints, which can be a serious failure source. The aft field joint, shown in Fig. 6.5, was the cause of the Challenger explosion.[41]

Liner is used at the grain/case interface to inhibit burning as the flame front arrives at the case wall. The liner composition is usually binder without added propellants. In areas of the case in which there is no grain interface, usually at the ends of the motor, insulation is used to protect the case for the full duration of a firing.

Grain Shape

Grain shape, primarily cross section, determines the surface burning area as a function of time. Burning area, along with burning rate, determines thrust. For a given propellant, surface area vs time determines the shape of the thrust-time curve. A cylindrical grain inhibited on the sides, a "cigarette burner" or "end burner," would have a constant burning area and a constant thrust. A grain with an

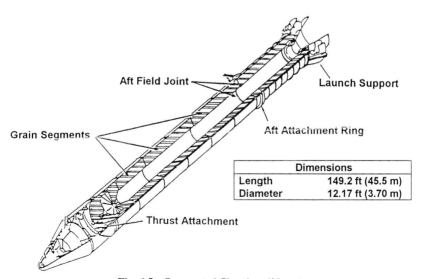

Fig. 6.5   Segmented Shuttle solid motor.

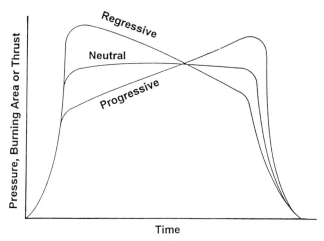

Fig. 6.6   Progressive, regressive, and neutral burning.

essentially constant burning surface area and thrust-time curve is called *neutral burning*. End-burning grains were used by the thousands for JATO rockets in the 1940s.

A grain that has an increasing surface area with time is called *progressive*. A case-bonded grain with a cylindrical port would be a progressive burning grain (see Fig. 6.6). The primary disadvantage of progressive burning is that, as the mass decreases, the thrust increases. The resulting loads on a spacecraft increase sharply with time, which is not conducive to an efficient structure. A cylindrical grain inhibited at the ends and burning from the sides would be *regressive*. *Neutral burning* is the most common design for spacecraft. Grain cross sections are shown in Fig. 6.7.

Note that a crack in the grain provides an unplanned increase in exposed area with a resultant sudden increase in pressure, which can lead to failure. Numerous precautions must be taken to prevent grain cracks, including:

1) Limits on grain temperature extremes.
2) Limits on shock loads on the grain and bond line.
3) A final set of grain x rays at the launch site just before launch.

A star grain is the most commonly used shape and is nearly neutral. The progression of the burning surface in a star grain is as shown in Fig. 6.8. Note that the flame front does not reach the liner in all locations at the same time. When the flame front first reaches the liner, chamber pressure starts to decay. At a given pressure, aggressive burning, or deflagration, can no longer occur. The residual, usable propellant that remains when combustion stops is called *sliver*. In a well-designed motor, sliver will be less than 2% of the grain weight. The chamber pressure and thrust as a function of time for a star grain is shown in Fig. 6.9.

Shelf Life

The shelf life of a motor is a conservative estimate of the acceptable length of time from the pour date to the firing date. Given that a typical shelf life is three

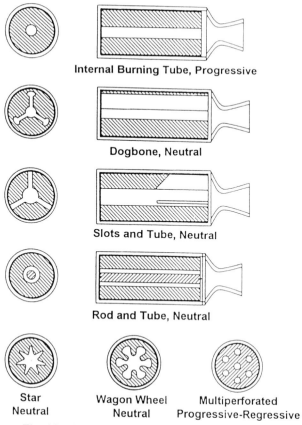

Internal Burning Tube, Progressive

Dogbone, Neutral

Slots and Tube, Neutral

Rod and Tube, Neutral

Star
Neutral

Wagon Wheel
Neutral

Multiperforated
Progressive-Regressive

Fig. 6.7   Grain configurations (from Ref. 7, p. 278).

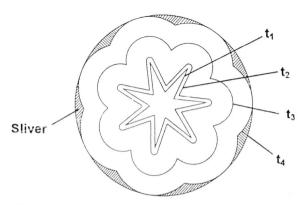

t_1

t_2

t_3

t_4

Sliver

Fig. 6.8   Burning surface progression in a star grain (Reprinted with permission from H. Koelle, *Handbook of Astronautical Engineering*, McGraw-Hill, 1961.).

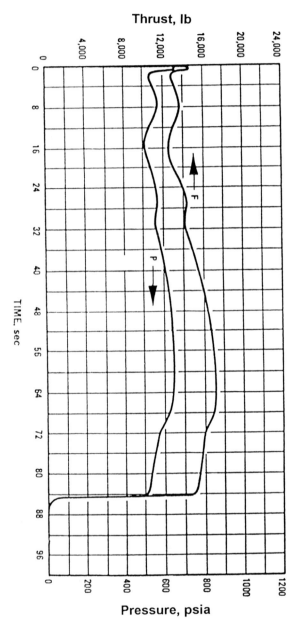

Fig. 6.9   Thrust-time curve for a star grain. (Courtesy Morton-Thiokol.)

years, the pour date must not be more than three years from the on-mission firing. Thus, the pour date becomes a major program milestone.

Off-Load Capability

Many solid motors have been demonstrated to perform properly over a range of propellant loads. This is called *off-load capability* and is useful in matching the required total impulse with the available total impulse. Off-loading is accomplished by pouring the motor case less than full. The Star 48, for example, is normally loaded with 4402 lb of propellant. Propellant can be off-loaded down to 3544 lb.

6.3   Thrust Control

A spacecraft must make provision to start and terminate thrust as well as to align the nominal vector through the center of mass and move the vector through small angles for vehicle stability.

Ignition

During most of a spacecraft assembly and flight, the objective is to prevent solid motor ignition. At those times when people are around the motor, it is a common requirement for the ignition system to be two fault–tolerant. The system must be designed such that unwanted ignition cannot occur after any two conceivable failures. If the spacecraft is launched on the Space Shuttle, this requirement is in effect until deployment. When impulse is required in the mission, it is common to require single-fault tolerance.

A typical solid rocket motor ignition system is shown schematically in Fig. 6.10. It requires careful use of redundancy to meet the failure criteria. Note that redundancy is not shown in Fig. 6.10 in the interest of simplicity. A typical ignition system consists of:

1) Squibs: Cylinders that contain a small chemical charge so sensitive that the heat from an electrical filament will cause a substantial energy release. The squibs have an electrical connector that receives the start signal. Squibs are required to have a specified all-fire current and a specified no-fire current. The electrical filament is called *a bridge wire*. Some squibs have redundant bridge wires to receive redundant signals.

2) Explosive transfer assemblies (ETA): An explosive train that runs from the S&A to the igniter. An ETA has the same function as a dynamite fuse but is much, much faster.

3) Safe and arm (S&A): A dc motor that houses the squibs on one side and the ETAs on the other, with a rotating cylinder between. A section of the explosive train is embedded in the rotating cylinder. The flame front that starts in the squibs must pass through the cylinder to reach the ETAs. When the cylinder is rotated to the safe position as shown in Fig. 6.10, the train is interrupted. In the armed position, the train is continuous; the cylinder is rotated 90 deg from the position shown in Fig. 6.10. There is a physical flag on the exterior of the S&A, which shows position. In addition, a position signal is sent to the command and data system so that the position can be telemetered to the ground operations team. It is typical to use dual S&As and dual igniters.

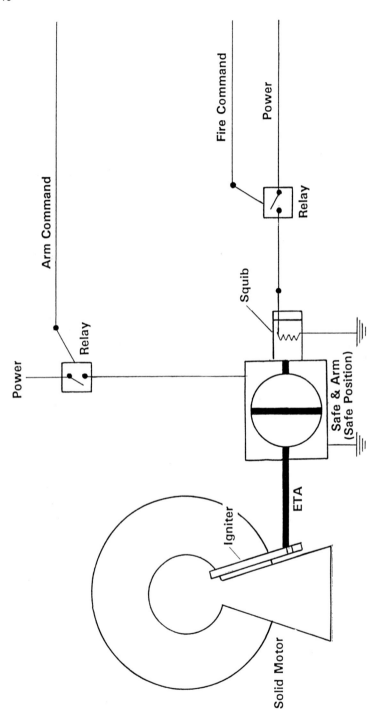

Fig. 6.10   Solid motor ignition system.

4) Igniter: Similar to a small solid motor with a propellant mixture that is readily ignited. The igniter raises the pressure in the motor port and raises the grain temperature to the ignition point.

At the appointed time, the spacecraft command system sends the "arm" command to a relay that powers the S&A motor. (In the Magellan mission, this point was 15 months after launch.) Hopefully, the S&A signals that the armed position has been achieved.

The "fire" command is sent to a separate relay, which powers the squib bridge wires, lighting the explosives in the squib. The flame front progresses through the S&A and the ETA and lights the motor igniters. The igniters bathe the grain surface in flame, and the motor ignites. The time from signal to grain ignition is about 150 ms; full thrust takes about another 50 ms.

Thrust Vector Control

Moving the net thrust vector through small angles for vehicle stability and maneuvering is called *thrust vector control* (TVC). The methods in common use are summarized in Table 6.2. All of the thrust vector control schemes in Table 6.2 except jet vanes require auxiliary propulsion to achieve roll control. All types give pitch and yaw control with a single nozzle.

Jet vanes and jetavators operate with low actuator power and provide high slew rates; however, there is a thrust and I_{sp} loss of up to 2%. Jet vanes, jetavators, and movable nozzles, TVC types 1–4 in Table 6.2, all require an actuator system. Hydraulic and electromechanical actuator systems are used. The IUS uses an electromechanical system consisting of:

1) A controller, which receives analog pitch and yaw commands from the guidance system. The analog signals are converted into pulsewidth–modulated voltages for the actuator motors.

2) Actuator control motors (one pitch, one yaw), which drive the ball screw mechanisms.

3) A ball screw mechanism, which positions the nozzle (one pitch, one yaw). The most common TVC methods are movable nozzles and fluid injection, types 3, 4, and 5 in Table 6.2. Figure 6.11 shows the Shuttle solid motor nozzle incorporating gimbal movement with a flexible bearing nozzle. A flexible bearing nozzle has no hot moving parts and provides repeatable, albeit high, actuator forces.

Table 6.2 Common methods of thrust vector control

Type	Flight use
1) Jet vanes	Jupiter, Juno
2) Jetavators	Polaris, Bomarc
3) Ball and socket nozzle	Minuteman
4) Flexible bearing nozzle	IUS, Shuttle
5) Liquid injection	Titan III, IV, Minuteman
6) Auxillary propulsion	Magellan

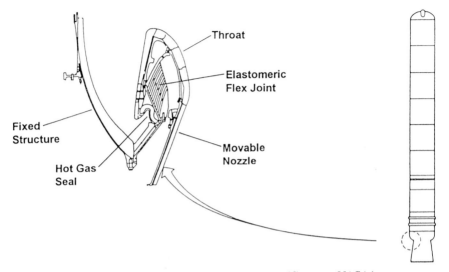

Fig. 6.11   Flexible bearing gimbaling nozzle. (Courtesy NASA.)

The fluid-injection systems inject a liquid into the nozzle, which causes an oblique shock in the exhaust stream, deflecting the thrust vector as shown in Fig. 6.12. The thrust vector is diverted in the direction opposite the injection point. The advantages of a liquid-injection system are high slew rate and increased motor performance. The disadvantages are the additional equipment and the liquid propellant required. The Titan family solid rocket motors use liquid injection, with nitrogen tetroxide used as an injectant. The fluid system is shown in Fig. 6.13.

The pyroseal valve shown in Fig. 6.13 seals the injectant prior to ignition. At ignition, the exhaust melts the exposed end of the valve, allowing injectant flow.

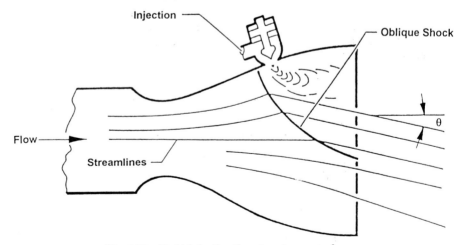

Fig. 6.12   Fluid-injection thrust vector control.

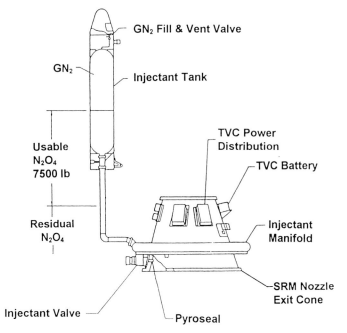

Fig. 6.13 Titan fluid-injection thrust vector control system. (Courtesy UTC and Lockheed Martin.)

An alternative approach to controlling the motor thrust vector is to provide an auxiliary propulsion system to stabilize the spacecraft. This was the approach used by Magellan. The monopropellant attitude-control system was designed to provide three-axis control with or without the solid motor firing. Redundant, aft-facing 100-lb thrusters provided pitch and yaw control during the firing. The system also provided vernier impulse adjustment after solid motor shutdown.

Thrust Vector Alignment

For three axis–stabilized spacecraft, the solid motor must be installed with the thrust vector accurately aligned with the spacecraft center of mass (c.m.). The severity of this requirement depends on the magnitude of attitude-control authority, the accuracy of c.g. location, and the accuracy of the thrust vector location. For a geometrically perfect nozzle, the thrust vector lies on the centerline. The thrust vector is very nearly on the centerline in real nozzles as shown in Table 6.3, for the Star 48B solid motor. The shape and size of the nozzle changes during the burn. The resulting variation in thrust vector location is more difficult to determine. The c.m. of a motor moves during the burning process; the tolerance about nominal movement is also difficult to determine.

In the Magellan spacecraft design, it was necessary to use 8 lb of ballast to bring the c.m. to within the required 0.02 in. of the predicted thrust vector location. It is difficult to obtain an actual spacecraft c.m. It must be done late in the process at the launch site and, even then, certain equipment will not be installed; for example,

Table 6.3   Star 48B nozzle alignment data (from Ref. 29, p. 9)

	Radial offset, in.	Angular error, deg
Population range	0.003–0.11	0.004–0.059
Average	0.0072	0.0247
One sigma	0.0026	0.0152
Magellan motor	0.0085	0.01761

liquid propellants will not be loaded, the solid motor will be represented by an inert motors, igniters will not be installed, and remove-before-flight equipment will still be installed. A spin-stabilized spacecraft avoids this set of problems because rotation nullifies thrust vector alignment errors.

Thrust Termination

Thrust termination is desirable because the delivered impulse cannot be predicted exactly. Delivered impulse depends on the temperature of the grain, the actual propellant weight, the exact composition, and the weight of inert parts consumed. Thrust termination allows a spacecraft to measure velocity gained and shutdown when the desired velocity is reached. The impulse uncertainty with thrust termination is reduced to the shutdown impulse uncertainty. Table 6.4 lists demonstrated thrust termination methods. Thrust can be terminated by suddenly reducing chamber pressure below certain limits, quenching the combustion, or nullifying thrust. Titan, for example, terminates stage 0 thrust by blowing out forward ports. Magellan and the Shuttle are designed not to require thrust termination. In both these cases, the solid motor burns out before the desired impulse is achieved. The deficit is made up by liquid engines that can be readily shut-down on command.

6.4   Cases

Materials

Table 6.5 shows the prominant solid motor case materials and their properties. Maraging steels are an alloy containing nickel, cobalt, and molybdenum. They are an attractive option because they can be machined in the anneald condition and can be aged at relatively low temperatures to reach high strength. (The steels are

Table 6.4   Thrust termination techniques

Action	Result
Venting forward ports	Balance thrust/reduce pressure
Chamber destruction	Reduced chamber pressure
Liquid quenching	Extinguish flame
Nozzle ejection	Reduced chamber pressure

Table 6.5   Solid motor case materials

	Titanium 6AL-4V forging	4130 steel	17-7 PH stainless steel	Maraging steel	D6aC steel	Filament-wound composites
Tensile strength, F_{tu}, ksi	130	180	180	200	220	150–250
Young's Modulus, 10^3ksi	16	29	29	28	29	4–11
Density, lb/in³	0.160	0.283	0.276	0.289	0.283	0.05–0.07
Temperature limit, °F[a]		575			900	

[a]Temperature limit for the ultimate tensile strength shown. [b]Data from MIL-HDBK-5. Data for comparison only; for design, consult the latest MIL-HDBK for the specific material, condition, and heat treatment.

aged in the martensitic condition, hence, maraging steels.) Titanium is the most common material for spacecraft motors; however, very rapid progress is being made with composites.

Mounting

There are myriad mounting provisions for solid rocket motors. Spacecraft loads may be carried through the case (IUS motor) or carried around the motor (Magellan). Mounting pads (Space Shuttle) or mounting rings (IUS) may be used. Mounting pads carry a point load, and an interface truss is desirable. Mounting rings carry a distributed load, and an interfacing shell structure is desirable. For large spacecraft motors, the dominant mounting design is a mounting ring. Figure 6.14 shows the Magellan SRM adapter structure designed to mount the Star 48B. The SRM adapter structure attaches at the forward mounting ring of the Star 48B. The adapter is a conical aluminum honeycomb structure that mates with a truss the Magellan central structure.

Jettison

After burnout, it is desirable to jettison the spent motor case and as much support structure as possible in order to reduce vehicle mass. Reduced mass reduces the energy required by the attitude-control system and the energy required for any subsequent maneuvering. There are two basic system types used for jettisoning: 1) linear-shaped charge and 2) explosive bolts or nuts. Linear-shaped charges are used to cut all the way around the girth of a shell structure. The freed motor case and attached structure are pushed away by springs or by gas from the linear charge.

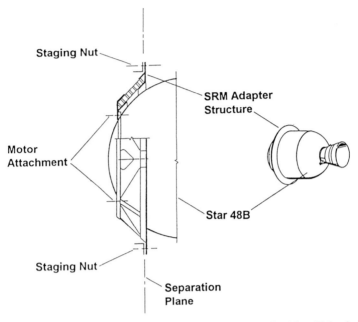

Fig. 6.14   Magellan SRM adapter structure. (Courtesy Lockheed Martin.)

Explosive nuts (or bolts) release a bolted joint in the structure, releasing the spent motor and support structure. The Magellan design used explosive nuts. In Fig. 6.14, the separation plane is between the main structure and the SRM adapter. This joint is held together by explosive bolts. When the nuts are blown, the spent Star 48B and the SRM adapter structure are pushed away with springs. The resultant reduction in spacecraft mass is 360 lb.

6.5 Nozzles

A solid rocket nozzle is the hottest part of the motor, especially at the throat. The design of a solid rocket nozzle is governed by the temperatures, erosion, and ablation of the parts. Note that erosion of the throat results in performance variation as a function of time. Figure 6.15 shows the elements of a typical nozzle and their functions. The nozzle inlet area becomes larger and more complicated if the nozzle is submerged.

Synthetic graphite is a low-strength refractory material commonly used as a throat insert. Structural support is required for graphite inserts. Carbon-carbon is a composite material consisting of carbon fiber material in a pyrolized carbon matrix. The carbon fibers provide outstanding specific strength and modulus. Carbon and silica phenolics are also composite materials.

6.6 Performance

Solid motor performance calculation must be approached somewhat differently than liquid system performance, as detailed in this section.

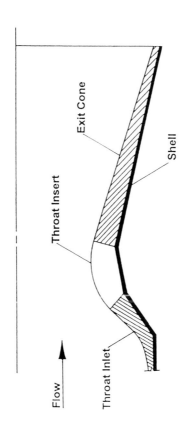

Element	Common Materials
Shell	4130 Steel 6 AL–4V Titanium Composites
Throat Inlet	Graphite Carbon Phenolic
Throat Insert	Graphite 3-D Carbon-Carbon Graphite Washers Forged Tungsten
Exit Cone	Carbon Phenolic Silica Phenolic Graphite Carbon/Carbon

Fig. 6.15   Typical solid rocket nozzle.

Burning Time and Action Time

Burning time and action time for a solid rocket motor can be estimated from internal ballistics using strand burning rates or can be measured on a test stand. As shown in Fig. 6.16, two times are of interest: 1) the burning time, which is measured starting from 10% thrust at ignition to 90% thrust during shutdown, and 2) the action time, which is measured starting at 10% thrust during ignition to 10% thrust during shutdown. (The thrust level from which the 10% and 90% points are calculated is the maximum thrust level.) The 90% thrust point on shutdown is the approximate time when the flame front reaches the inhibitor at the case wall. From the 90% point down to the 10% point, the slivers are being consumed (less efficiently than the combustion at full chamber pressure).

For a motor firing in a vacuum, thrust will continue for a substantial period of time after the 10% thrust point on shutdown. During this *tail-off period*, a low chamber pressure is maintained by burning sliver, outgassing inert materials, and low-level smoldering of hot insulation, nozzle materials, and liner. The impulse, thrust level, and time to zero thrust are nearly impossible to measure in the atmosphere because a hard vacuum cannot be maintained with a motor outgassing. It is very important in the design of the spent case staging sequence to provide an adequate period before release in order to make sure the case will not follow the spacecraft.

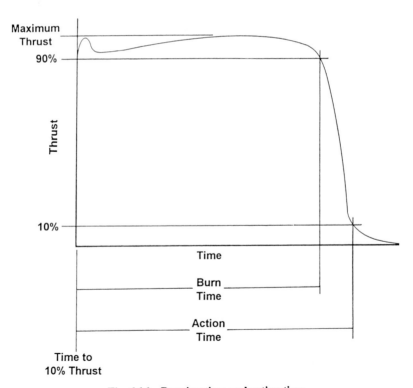

Fig. 6.16   Burning time and action time.

Table 6.6 Effective propellant weight

	Star 48B	Minuteman Stage I
Initial weight, lb	4,660	50,550
Burnout weight, lb	235	4,264
Effective propellant, lb	4,425	46,286
Loaded propellant, lb	4,402	45,831
Inerts consumed, lb	23	455

Effective Propellant Weight

The *effective propellant weight* is the propellant actually consumed by the motor. Effective propellant is the difference between the initial total weight of the motor and the postburn total weight of the motor.

$$W_{\text{eff}} = W_i - W_f \qquad (6.6)$$

Effective propellant weight can be either more or less than the total propellant load. There are two effects at work: 1) sliver, which tends to reduce effective propellant, and 2) consumed inert parts, which tend to increase effective propellant. Consumed inert weights are nozzle materials that ablate, particularly in the throat area, as well as volatiles forced out of the insulation and liner. In a vacuum motor, it is very likely that the sliver is consumed during the shutdown or tail-off; thus, spacecraft motors usually have an effective weight greater than the total propellant loaded. Table 6.6 shows two examples.

Impulse

Two impulse values result from the two thrust times described. If thrust is integrated from time zero to time $= t_b$, the resulting impulse, I_b, is called *burning time impulse*:

$$I_b = \int_0^{t_b} F \, dt \qquad (6.7)$$

If action time is used, I_a, action time impulse, results:

$$I_a = \int_0^{t_a} F \, dt \qquad (6.8)$$

Action impulse is often called *total impulse*, which is true in the sense that the tail-off impulse is essentially unusable. However, the tail-off impulse must be provided for as already noted.

Thrust

Two versions of time average thrust arise from the dual time standard discussed:

$$\bar{F}_b = \frac{I_b}{t_b} \tag{6.9}$$

$$\bar{F}_a = \frac{I_a}{t_a} \tag{6.10}$$

where

$\bar{F}_b$ = burn time average thrust, lb
$\bar{F}_a$ = action time average thrust, lb
t_b = burning time, s
t_a = Action time, s
I_b = burning time impulse, lb-s
I_a = action time impulse, lb-s

The burn time average thrust $\bar{F}_b$ the most often used and is frequently referred simply (and imprecisely) as *average thrust*. For the Star 48:

Burn time average thrust = 15,000 lb
Action time average thrust = 14,980 lb

Maximum thrust is critical to the design of the case. It is calculated from internal ballistics for a hot grain using maximum burning rate.

Specific Impulse

Theoretical specific impulse for the propellant may be calculated from theoretical C^* and C_f as follows:

$$I_{sp} = \frac{C^* C_f}{g_c} \tag{6.11}$$

C^* and C_f can be obtained from propellant thermodynamic properties as discussed in Chapter 2. Real motors get a C^* efficiency of about 93% and a C_f efficiency of about 98%.

Propellant I_{sp} can also be obtained from thrust-time test data because

$$I_{sp} = \frac{I_a}{w_p} \tag{6.12}$$

where

I_{sp} = propellant specific impulse, s
w_p = propellant weight loaded, lb

In calculating propellant I_{sp} from Eq. (6.12), the loaded propellant weight is used. The effective specific impulse is computed from effective propellant weight:

$$I_{eff} = \frac{I_a}{w_i - w_f} \tag{6.13}$$

where

I_{eff} = effective specific impulse, s
w_i = initial motor weight (total), lb
w_f = final motor weight after burnout, lb

Motor Weight

The total weight of a solid rocket motor can be expressed as

$$W = \frac{I}{\eta I_{\text{sp}}} \qquad (6.14)$$

where

W = total motor weight, lb
I = total (action time) impulse, lb/s
I_{sp} = propellant specific impulse, s
η = propellant mass fraction, or W_p/W_i

The total weight of 23 spacecraft solid rocket motors of current design is shown in Fig. 6.17 as a function of total impulse. For these motors, the average propellant I_{sp} is 290, and the average propellant mass fraction is 0.93. Equation (6.15) fits Fig. 6.17 with a correlation coefficient of 0.954:

$$W = 0.0037I \qquad (6.15)$$

Equation (6.16) satisfies both Eqs. (6.13) and (6.14) when $I_{\text{sp}} = 290$ and $\eta = 0.93$:

$$W = \frac{I}{0.93 I_{\text{sp}}} \qquad (6.16)$$

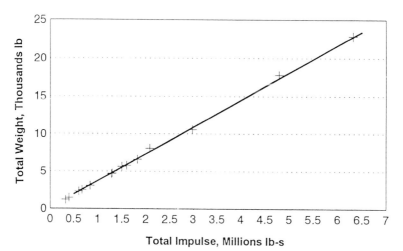

Fig. 6.17   Solid rocket motor weight.

Example 6.2: Solid Motor Performance Size and Mass

Empirical estimates of motor performance, size, and weight are very useful, particularly in preliminary design phases. The following example shows how this can be done.

Design a solid rocket motor to meet the following requirements:

Total (action time) impulse $= 3.709 \times 10^5$ lb-s

Average (action time) thrust $= 7140$ lb

It is necessary to make a few additional assumptions so that the design can proceed. Assume:

Motor shape: spherical

Nozzle: 15-deg half-angle cone, 25% submerged

Average chamber pressure $= 560$ psia

Area ratio $= 35$

Propellant: HTPB/AP/AL

These assumptions can be revisited in later, second-order trade studies.

Action Time

Motor action time is

$$t_a = \frac{(3.709E5)}{7140} = 51.95 \text{ s}$$

Theoretical C_f and C^*

Theoretical C_f and C^* can be calculated from the thermodynamic properties of the propellant HTPB/AP/AL. From Appendix B:

Molecular weight $= 22$

Ratio of specific heats $= 1.26$

Combustion temperature $= 6160°R$

Density $= 0.0651$ lb/in³

Burning rate $= 0.282$ in./s

Calculating theoretical C^* yields

$$C^* = \frac{\sqrt{k g_c R T_c}}{k\sqrt{\left(\frac{2}{k+1}\right)^{\frac{k+1}{k-1}}}} \tag{6.17}$$

$$C^* = \frac{\sqrt{(1.26)(32.17)(1544/22)(6160)}}{(1.26)\sqrt{\left(\frac{2}{2.26}\right)^{\frac{2.26}{0.26}}}}$$

$$C^* = 5650 \text{ fps}$$

Actual C^* can be estimated as 93% of theoretical, or 5254 fps.

Calculating theoretical C_f yields

$$C_f = \sqrt{\frac{2k^2}{k-1}\left(\frac{2}{k+1}\right)^{\frac{k+1}{k-1}}\left[1-\left(\frac{P_e}{P_c}\right)^{\frac{k-1}{k}}\right]} + \left(\frac{P_e-P_a}{P_c}\right)\frac{A_e}{A_t} \qquad (6.18)$$

The pressure ratio for an area ratio of 35 and $k = 1.26$ from Fig. 2.6 is 0.0022. Substituting into Eq. (2.33) yields

$$C_f = \sqrt{\frac{2(1.26)^2}{0.26}\left(\frac{2}{2.26}\right)^{\frac{2.26}{0.26}}\left[1-\left(0.0022^{\frac{0.26}{1.26}}\right)\right]} + (0.022)(35)$$
$$C_f = 1.817$$

Actual C_f can be estimated as 98% of theoretical, or 1.780.

Specific Impulse

Actual propellant specific impulse can be estimated as follows:

$$I_{sp} = \frac{(5254)(1.780)}{32.17} = 290.7 \text{ s}$$

Propellant and Motor Weight

If we estimate a mass fraction of 0.93, the total motor weight is

$$W = \frac{3.709E5}{0.93(290.7)} = 1375 \text{ lb}$$

The propellant weight is

$$W_p = \frac{3.709E5}{290.7} = 1275 \text{ lb}$$

Throat Area

Throat area corresponding to a chamber pressure of 560 psi is

$$A_t = \frac{\bar{F}}{P_c C_f} = \frac{7140}{(560)(1.780)} = 7.16 \text{ in}^2$$

Throat diameter is 3.02 in., exit area is 250 in², and exit diameter is 17.86 in.

Motor Volume

The motor volume can be obtained from the *volumetric loading fraction*, which is the ratio of the propellant volume to the motor volume, excluding the nozzle.

Using a volumetric loading fraction of 90%, which is typical for a spacecraft motor, yields

$$V_m = \frac{W_p}{\rho V_f} = \frac{1275}{(0.065)(0.90)} = 2175 \text{ in}^3$$

Motor Diameter

From motor volume, the motor diameter is

$$D_m = 2\left(\sqrt[3]{\frac{(0.75)(21795)}{\pi}}\right) = 34.6 \text{ in.}$$

Motor Length

The nozzle length, if a 15-deg half-angle cone is assumed, is

$$L_n = \frac{(17.86 - 3.02)}{2 \tan(15)} = 27.69 \text{ in.}$$

With the nozzle 25% submerged, the motor length is

$$L_m = 34.6 + (0.75)(27.69) = 55.4 \text{ in.}$$

Try this example using the PRO software.

6.7   Some Flight Motors

In this section, the characteristics of some solid motor designs, with extensive flight history, are characterized.

Explorer 1

The first U.S. spacecraft, Explorer 1, was also the first to carry a solid rocket motor. A small solid motor, integral with the spacecraft structure, inserted the spacecraft into orbit. The motor, shown in Fig. 6.18, was originally used as a scale model in the development of the Sergeant missile. Table 6.7 summarizes the

Table 6.7   Performance of the Explorer 1 solid motor
(from Ref. 42)

Parameter	Value
Effective burning time, s	5.35
Effective chamber pressure, psia	496
Maximum vacuum thrust, lb	2100
Total Impulse, lb-s	10566
Vacuum specific impulse, s	219.8

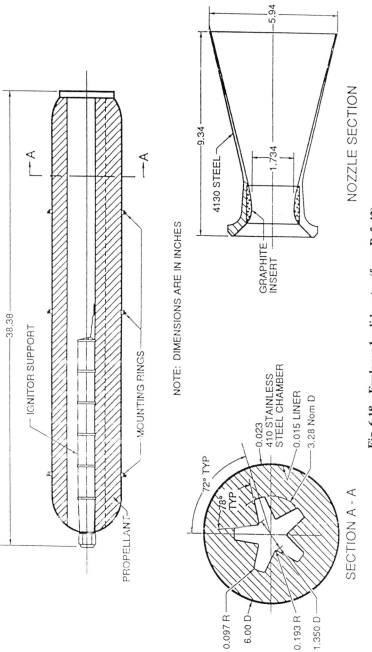

NOTE: DIMENSIONS ARE IN INCHES

NOZZLE SECTION

5.94

9.34

4130 STEEL

1.734

GRAPHITE INSERT

IGNITOR SUPPORT

38.38

MOUNTING RINGS

PROPELLANT

SECTION A - A

72° TYP

18° TYP

0.023

410 STAINLESS STEEL CHAMBER

0.015 LINER

3.28 Nom D

0.097 R

6.00 D

0.193 R

1.350 D

Fig. 6.18   Explorer 1 solid motor (from Ref. 42).

Table 6.8   Formulation of Explorer 1 propellant
(from Ref. 42)

Component	Weight percent
NH_4ClO_4 (oxidizer)	63.00
LP 33 (fuel)	33.17
Curing catalysts	3.5
Nylon tow (reinforcement)	0.33

performance of the motor. Research at JPL in the 1940s lead to the composites in use today. The polysulfide-based propellant used on Explorer 1 is an early example; the propellant formulation is shown in Table 6.8.

Explorer 1 was developed by the Jet Propulsion Laboratory in the explosive political atmosphere following the success of Sputnik. Explorer 1 was launched on a Jupiter C on January 31, 1958. Its primary scientific result was the discovery of the Van Allen belt.

Star 48B

The Star 48B, Fig. 6.19 and Table 6.9, was developed by Thiokol Corporation for use in the McDonnell Douglas Payload Assist Module. In addition, the Star 48B was chosen to put Magellan in orbit around Venus in 1990. The Star 48B contains 4431 lb of propellant and delivers 1.30 million lb-s of impulse. It has off-load capability down to 3833 lb of propellant.

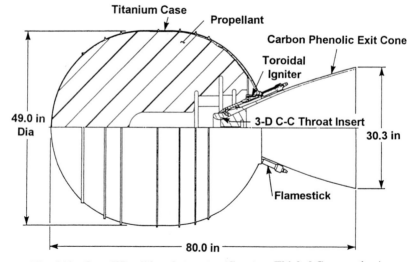

Fig. 6.19   Star 48B solid rocket motor (Courtesy Thiokol Corporation.)

Table 6.9 Star 48B motor characteristics (Courtesy Thiokol Corporation.)

Motor performance		*Weight, lb*	
tal impulse, lb-s	1,303,705	Total	4721
ximum thrust, lb	17,490	Propellant	4431
rn time ave. thrust, lb	15,430	Case assembly	129
tion time ave thrust, lb	15,370	Nozzle assembly	97
pellant specific impulse, s	294.2	Internal insulation	60
ective specific impulse, s	292.1	Liner	2
rn time/action time, s	84.1/85.2	SA, ETA	5
ition delay, s	0.099	Miscellaneous	3
rn time ave. chamber	579	Total inert	290
pressure, psia		Burnout	258
tion time ave. chamber	575	Propellant mass fraction	0.939
pressure, psia			
ximum chamber pressure, psia	618	*Propellant*	
		Propellant designation	TP-H-3340
Case		Formulation: Al, 18%	
terial	6Al-4V Titanium	AP, 71%	
nimum ultimate strength, ksi	165	HTPB binder, 11%	
nimum yield strength, ksi	155		
drostatic test pressure, psia	732	*Propellant configuration*	
nimum burst pressure, psia	860	Type	Internal burning,
minal thickness, in.	0.069		radial slotted star
		Web thickness, in.	20.47
Nozzle		Web fraction, %	84
it cone material	Carbon Phenolic	Sliver fraction	0
roat insert	2D Carbon/Carbon	Propellant volume, $in.^3$	68,050
ernal throat diameter, in.	3.98	Volumetric loading density, %	93.1
it diameter, in.	29.5	Web ave. burning surface	3325
pansion ratio, initial/average	54.8/47.2	area, $in.^2$	
it cone half-angle, exit/eff.,	14.3/16.3	Initial surface to throat area ratio, K	226
deg			
pe	Fixed, Contoured	*Propellant characteristics*	
		Burn rate @ 1000 psi, in./s	0.228
Liner		Burn rate exponent	0.30
pe	TL-H-318	Density, $lb/in.^3$	0.0651
nsity, $lb/in.^3$	0.038	Temperature coefficient	0.10
		of pressure	
Igniter		Characteristic velocity, ft/s	5010
in. firing current, amp	5.0	Adiabatic flame temperature, °F	6113
rcuit resistance, ohm	1.1	Ratio of specific heats: chamber	1.14
. of detonators and TBIs	2	Nozzle exit	1.18

Table 6.10   IUS solid rocket motors (Courtesy Boeing Aerospace Company, Ref. 26)

	Stage I	Stage II
Motor total weight, lb	22,981	6,633
Total impulse, lb-s	6,324,882	1,752,235
Total impulse, (with EEC)	—	1,839,514
Nominal propellant load, lb	21,404	6,061
Minimum propellant load, lb	10,700	3,000
Burn time (no off-load), s	152.02	103.35
Average thrust, lb	41,611	16,954
Average thrust (with EEC), lb	—	17,799
Average chamber pressure, psi	579	610
Maximum chamber pressure, psi	817	862
Delivered I_{sp} (without EEC), s	295.5	289.1
Delivered I_{sp} (with EEC), s	—	303.5
Throat diameter, in.	6.48	4.2
Area ratio	63.80	49.33
Area ratio (with EEC)	—	181.60
Exit diameter (without EEC), in.	51.76	29.5
Exit diameter (with EEC), in.	—	56.6
Motor outside diameter, in.	92.0	63.3
Overall length (without EEC), in.	123.98	66.5
Gimbal movement, deg	4	7

Inertial Upper Stage

The inertial upper stage (IUS) motor was built by UTC for Boeing Aerospace Co. It is used with both the Space Shuttle and the Titan family of launch vehicles. As shown in Fig. 6.20 and Table 6.10, the vehicle incorporates two solid rocket motors. Stage I contains 21,404 lb of propellant, and Stage II contains 6061 lb. Both stages have gimbaled nozzles operated by hydraulic actuators. The upper stage has an extendible nozzle. Both motors have carbon-carbon integral throats and two-dimensional carbon-carbon exit cones and redundant ignition systems.

The grains have a tubular design that allows 50% off-load capability. The motors are cast on a mandrel and then machined to nominal or off-load requirements. Off-load machining follows the grain burning pattern. This approach does not alter the thrust profile. The igniter can accommodate 50% off-loading. Roll control for both motors is provided by a monopropellant reaction-control system on the second stage. (The system provides three-axis control when the motors are not firing.) Stage II can be flown with or without its extendible exit cone (EEC). The increase in area ratio provides an additional 14.4 s in specific impulse.[26]

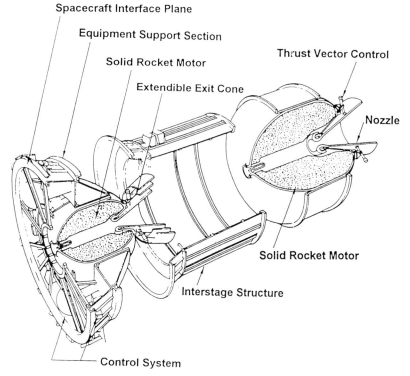

Fig. 6.20   Inertial upper stage (from Ref. 26, p. 9).

Problems

6.1   Consider a cylindrical motor 18 in. i.d. and 48 in. long, with a cylindrical port, a volumetric loading fraction of 93%, a mass fraction of 92%, a specific impulse of 290 s, and a PBAN/AP/AL grain inhibited at each end. Calculate:
  a) Port area
  b) Web thickness
  c) Burning time
  d) Propellant weight
  e) Loaded motor weight
  f) Thrust calculated at time 0, 1, 2, 3, and 4 s

6.2   Determine the burning time, action time, and maximum thrust from the solid motor test data shown in Fig. P6.2. By graphical integration, determine the action time impulse and the burning time impulse. What are the burning time average thrust and action time average thrust?

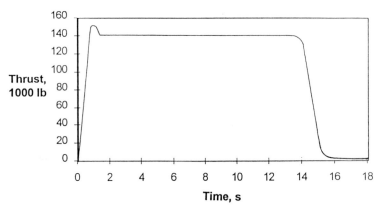

Fig. P6.2   Solid motor thrust-time curve.

6.3   Estimate a solid motor size, mass, vacuum specific impulse, vacuum C_f, C^*, action time, and throat area, given the following:
   Propellant: PBAA/AP/AL
   Motor shape: spherical
   Nozzle: 15-deg half-angle cone, 25% submerged
   Action time impulse = 1,008,000 lb-s
   Action time average thrust = 22,000 lb
   Average chamber pressure = 320 psi
   Area ratio = 28
Use a motor mass fraction of 86% and a volumetric loading fraction of 89%.

7
Cold-Gas Systems

Cold-gas systems are the simplest and oldest type used for attitude-control thrusters. In the 1960s era, cold gas was the most common type of system, and it is still used in cases in which the total impulse needed is less than about 1000 lb-s.

7.1 Design Considerations

Figure 7.1 shows a typical cold-gas system. A typical system includes the gas storage container, a gas-loading valve, filtration, a pressure regulation, pressure relief, and a series of thrusters with valves.

Gas is loaded through a ground fitting, V1, and filter F1 (typically 20 μ) into a titanium tank. (See Sec. 4.6 for tank design information.) Relatively high pressures are used, for example, 5000 psia. The gas may be isolated by an ordnance valve, V2, until release from the launch vehicle. Gas then flows through filter F2, which protects the regulator and flows at a reduced pressure through the low pressure filter F3, to the thruster valves, V3. On command, the thruster valves are opened in pairs to produce attitude-control torques. The small converging-diverging nozzles can be integral with the valve body. A relief valve protects the low-pressure system from regulator leakage or failure possibilities. The relief exhaust may be split to provide zero torque on the vehicle.

Candidate gases are shown in Table 7.1. Helium, nitrogen, and Freon-14 have flown. Although helium has the best performance and lowest gas weight, a leakage failure is less likely with nitrogen or Freon. From Table 7.1, note that the measured specific impulse is about 90% of theoretical, less than you would expect of larger thrusters. This lower performance is due primarily to the conical nozzles commonly used, as opposed to the contoured nozzles on larger thrusters.

Specific impulse can be calculated from ideal rocket thermodynamics,

$$I_{sp} = \sqrt{\frac{2kRT_c}{g_c(k-1)}\left[1-\left(\frac{P_e}{P_c}\right)^{\frac{k-1}{k}}\right]} \qquad (7.1)$$

Theoretical C_f is

$$C_f = \sqrt{\frac{2k^2}{k-1}\left(\frac{2}{k+1}\right)^{\frac{k+1}{k-1}}\left[1-\left(\frac{P_e}{P_c}\right)^{\frac{k-1}{k}}\right]} + \left(\frac{P_e-P_a}{P_c}\right)\frac{A_e}{A_t} \qquad (7.2)$$

Thrust is

$$F = P_c A_t C_f \qquad (7.3)$$

Table 7.1 Candidate gases

Gas	M	k	R	I_{sp} Theo.[b]	I_{sp} Meas.	Ref.[a]
Helium	4	1.659	386	176	158	25
Nitrogen	28	1.4	55.1	76	68	25
					67	44
Freon-14	88	1.22	17.6	49	46	24, 44
Freon-12	121	1.14	12.8	46	37	44
Ammonia	17	1.31	90.8	105	96	44
Hydrogen	2	1.40	767	290	260	44
Nitrous oxide	44	1.27	34.9	67	61	45

[a]Reference applies to the measured I_{sp} values. [b]Theoretical specific impulse is for vacuum, frozen equilibrium, area ratio $= 100$, gas temperature $= 560°$R.

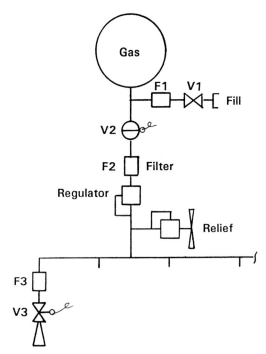

Fig. 7.1 Cold-gas system.

The gas weight and volume needed can be obtained from the total impulse requirement and the equation of state,

$$W_p = \frac{I}{I_{sp}}$$ (7.4)

and

$$V = \frac{W_p R T}{P}$$ (7.5)

Design of the tankage is discussed in Sec. 4.5.

Example 7.1: Cold-Gas System Design

Design a cold-gas system to meet the following requirements:
Total impulse = 1000 lb-s
Minimum impulse bit = 0.001 lb-s
Area ratio = 100
Thruster temperature maintained at 100°F
Use helium as a working fluid.

If a minimum valve response time of 20 milliseconds is assumed, the thrust level can be found as follows:

$$F = \frac{0.001}{0.02} = 0.05 \text{ lb}$$

A thrust level of 0.05 lb is within the capability of cold-gas thrusters. If the spacecraft is three axis–stabilized, 12 thrusters will be required to provide pure couples.

Obtain $P_e/P_a = 0.00008$ from Fig. 2.6 for an area ratio of 100; then, calculate theoretical specific impulse:

$$I_{sp} = \sqrt{\frac{2(1.659)(386)(560)}{(32.17)(0.659)}}\left[1 - (0.00008)^{\frac{0.659}{1.659}}\right] = 182 \text{ s}$$

Why is this value of theoretical specific impulse greater than that in Table 7.1? Actual specific impulse can be expected to be about 90% of theoretical or 164 s.

The helium weight required can be obtained from total impulse:

$$W = \frac{1000}{164} = 6.10 \text{ lb}$$

The tank volume needed to contain 6.10 lb of helium plus 0.5 lb of unusable helium is obtained from the equation of state,

$$V = \frac{(6.6)(386)(520)}{(3000)(144)} = 3.067 \text{ ft}^3$$

Why is a different temperature used in the equation of state and in the specific impulse calculation?

Tank design is covered in detail in Sec. 4.6. A spherical titanium tank, stressed to 100,000 psi maximum, for 6.6 lb of helium at a maximum pressure of 3000 psi, weighs 48.5 lb. Allowing 10 lb for valves, lines, filters, and thrusters, the cold-gas system weighs 65.1 lb.

7.2 Flight Systems

Viking Orbiter

The Viking Orbiter reaction-control system provided three-axis control of the spacecraft when the main engine was not firing. One of the two redundant systems shown in Fig. 7.2. The systems used nitrogen gas as the propellant. Each tank held 15.5 lb of nitrogen at an initial supply pressure of 4400 psig and a temperature of 100°F. The specific impulse was 68 s at 68°F. Nitrogen was filtered en route to six thrusters. The thrusters were mounted on the solar panel tips to maximize moment arm and were canted at 25 deg. to avoid impingement on the panels. The relief valve exhaust was branched to produce zero net torque.[25]

The system was designed to operate on helium borrowed from the main propulsion system in the event of failure. The system performance using nitrogen and helium is summarized in Table 7.2.[25] The Viking Orbiter gives a rare opportunity to compare the performance of two gases in the same thrusters. The minimum impulse bit was not measured with helium; however, from Eq. (4.2), I_{min} is estimated to be 0.00061 lb-s for helium.

LANDSAT 3

The LANDSAT 3 was the third of a series of Earth resource surveyor spacecraft. The 2100-lb spacecraft provided complete multispectral images of the Earth's surface every 18 days. The attitude-control system provided three-axis stabilization and nadir pointing for the spacecraft. Attitude-control torques were provided by six Freon-14 thrusters. The 480-in.³ propellant tank was initially charged with 12.2 lb of Freon-14 at 2000 psia at 20°C. Freon was fed to the thrusters through a pressure regulator. Each thruster provided about 0.1 lb of thrust. The system provided a total impulse of 563 lb-s at a measured specific impulse of 46.2 s.[24]

Table 7.2 Viking Orbiter performance

	Nitrogen (primary mode)	Helium (backup mode)
Thrust per jet, lb	0.030	0.0278
Specific impulse @ 68°F, s	68	158
Minimum impulse bit @ 22 ms, lb-s	0.00063	

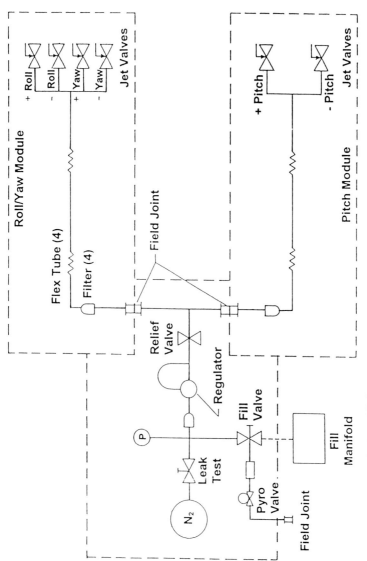

Fig. 7.2 Viking Orbiter reaction-control system (from Ref. 25, p. 189).

Appendix A
PRO: AIAA Propulsion Design Software

A.1 Introduction to PRO

PRO: AIAA Propulsion Design Software is a collection of fundamental tools that are versatile enough to accommodate a wide range of situations and powerful enough to save a lot of work. PRO is designed for the feasibility study phase of a spacecraft project, with emphasis on sizing and preliminary performance calculations.

A.2 Quick Tour of PRO: AIAA Propulsion Design Software

In this section, we make a quick tour through the features, arrangement, and conventions used in PRO. More detail can be obtained in subsequent sections.

Installation

PRO requires no special installation. If you need any help with hard drive installation, see Sec. A.11. If you are experienced, just prepare a directory and use the COPY *.* command to copy all the files on the master disk into the directory.

To start

Go to the hard disk directory where you installed PRO, or put the PRO disk in drive A.

Type PRO (caps or lower case) and press Enter.

When the title screen appears, PRO is ready to use.

Menus

PRO is menu-driven. The Main Menu allows you to select the PRO function you want from the following list:

1) *Propulsion Requirements* calculates the propulsion requirements for performing various orbital and attitude-control maneuvers.

2) *Rocket Engine Design* allows you to estimate the performance parameters, size, and weight of monopropellant, bipropellant, cold-gas, and solid rocket motors.

3) *Pulsing Engine Performance* estimates the performance to be expected from an engine, given the characteristics of the pulsing duty cycle.

4) *Blowdown System Performance* calculates the pulsing or continuous performance of a blowdown propulsion system as tank pressure decreases.

5) *System Weight Statement* is a spreadsheet for tabulating system weight.

6) *Utilities* makes theoretical performance, equation of state, tank design, and propellant-control device design calculations as well as unit conversions.

7) *Quit* exits PRO and returns to DOS.

Selection Menus merely require that you indicate what you want to do. The Main Menu and the Utilities Menu are examples of this type. To make a selection, type the number of the menu item or move the cursor to the item and press Enter.

To move the cursor among the selections, use:

1) Up and Down Arrows from the numeric keypad or the extended keypad.

2) Page Up and Page Down from either keypad.

3) Home or End from either keypad.

The cursor will wrap from the last menu item to the first, and vice versa.

Input Menus allow you to enter the parameters for a given calculation. For example, the Pulsing Performance Menu lists eight parameters necessary for estimating engine performance during pulsing. To do this:

1) Move the cursor to an element you want to enter.

2) Type your entry. To edit an error, use the Left Arrow or the Backspace key; the cursor will erase as it moves left. You may use exponential input if you wish, for example, 1.23456E6.

3) When you are satisfied with your input, press Enter, Tab, Up Arrow, or Down Arrow.

As soon as the requisite number of items are entered, PRO erases the menu and proceeds with the calculation. You may move up and down the menu list, editing entries, until PRO senses a value for each needed parameter, eight parameters in this case.

Some input menus offer a choice of equivalent parameters to enter. For example, the Equation of State Menu asks for gas pressure and temperature plus either volume *or* weight. In this case, the Input Menu provides for four entries but needs only three parameters. As soon as the third parameter is entered, PRO proceeds with calculation. In cases in which alternative entries are acceptable, a note below the menu list announces the acceptable choice, for example:

Input Volume *or* Weight

Two of the input menus, Weight Statement and Start System File, do not proceed with calculation until you signal that you are satisfied with all entries. Alt-C, for Continue, is the command to proceed. The operative commands for each menu are listed at the bottom of the screen. Numerical inputs are screened to eliminate unwanted characters such as commas, percent signs, and letters other than *E*.

Esc

The Escape key takes you back one menu at a time to the Main Menu. The Escape key is active almost anywhere in PRO>.

Units

PRO expects inputs in English system of units:
Thrust, pounds
Flow rate, pounds/second

Specific impulse, seconds
Angles, degrees
Mass, slugs
The output on screen and printout is in the same units although the units are abbreviated. The Utility Menu provides for rapid and convenient conversions.

Physical Properties

The physical properties used by PRO are tabulated in Appendix B and can be changed by using the SETPROP program.

Equations Used

The equations used in PRO are all developed and referenced to source in the main text. References and discussion are not repeated in this manual.

Printed Reports

The last line in any result screen asks if you want a printed report. A yes response (just press Enter) will print results to a wide variety of printers. You will be asked to enter the heading you want; this input is optional. If a report is given a heading, it is automatically dated in the last heading line. Type N or n to indicate that no report is desired.

A.3   The Main Menu

PRO works from a Main Menu, which allows you to direct its operations. The menu, shown in Fig. A.1, is the first screen shown when PRO is started. At the top of the screen, after the title, is the version number. At the bottom of the screen are abbreviated directions for selection from the menu. The menu functions may be

```
                  PRO:AIAA Propulsion Design Software
                          by Charles D. Brown

 (C)Copyright 1995. American Institute of Aeronautics and Astronautics.
                          All Rights Reserved


              ► 1 PROPULSION REQUIREMENTS
                2 ROCKET ENGINE DESIGN
                3 PULSING ENGINE PERFORMANCE
                4 BLOWDOWN SYSTEM PERFORMANCE
                5 SYSTEM WEIGHT STATEMENT
                6 UTILITIES
                7 QUIT






      V 3.0   Press Menu No. to Select  -OR-  ↑↓ to Highlight Then Enter
```

Fig. A.1   Main Menu.

selected by simply typing the menu number or by using the arrow keys to highlight the desired item and pressing Enter. Seven selections are available from the Main Menu:

1) *Propulsion requirements.*   This function performs the calculations described in detail in Chapter 3. First, the type of maneuver is selected: orbital maneuver, one-axis maneuver, one-sided limit cycle, spin-axis precession, or reaction-wheel unloading. Then, the propellant required and other particulars of the maneuver are calculated.

2) *Rocket engine design.*   First, the propellant is selected; then, the engine parameters are entered: thrust, area ratio, and chamber pressure. PRO then calculates engine performance, weight, and dimensions.

3) *Pulsing engine performance.*   Given the steady-state performance of an engine, the pulse width, the engine offtime, and prefiring temperature of the chamber, PRO will calculate the pulsing specific impulse.

4) *Blowdown system performance.*   First, a system file is prepared containing the performance parameters of a given system. Then, PRO will perform the calculations involved in defining performance as a function of time as tank pressure decays. Pulsing or continuous operation may be chosen.

5) *System weight statement.*   PRO provides a spreadsheet for the continuous update of system weight.

6) *Utilities.*   The utilities available are:
- Theoretical performance
- Equation of state
- Tank design
- Zero-*g* control device design
- Unit conversions

7) *Quit.*   This function returns you to DOS.
Subsequent sections discuss each function in turn.

A.4   Propulsion Requirements

The first order of business is to define the performance requirements for a propulsion system. Ultimately, these requirements can be reduced to a propellant weight requirement. The process is discussed in detail in Chapter 3. When you select this function from the Main Menu, you will be asked to select from the following types of maneuvers:

1) Orbital maneuvers (translational velocity change)
2) One-axis maneuvers
3) One-sided limit cycle
4) Spin-axis precession
5) Reaction-wheel unloading

Orbital Maneuvers

Orbital maneuvers consist of plane changes, Hohmann transfers, circularization, orbit trim, and the like. Each of these maneuvers requires one or more steady-state burns, producing a given velocity change. PRO uses the Tsiolkowski equations to compute the propellant weight required for a given maneuver.

When you select Orbital Maneuvers from the Requirements Menu, you will get a multiple choice input screen requesting velocity change, specific impulse, and

```
                         ORBITAL MANEUVERS

            Velocity Change, km/s  =  1.831
            Specific Impulse, s    =  290
            Initial Weight         =
            Final Weight           =  1500




                    Input any three.
            Weights can be in any unit system.




   Esc = Back Up    Move = ↑↓ PgUp PgDn    Edit = ←    Accept = Enter
```

Fig. A.2   Orbital Maneuvers Menu.

initial or final spacecraft weight. For example, a spacecraft requires a final velocity change of 1.831 km/s to achieve geosynchronous orbit. This velocity change is supplied by a solid rocket motor delivering a specific impulse of about 290 s. Assume that the spacecraft weight in geosynchronous orbit is 1500 lb. To use PRO to compute the usable propellant for the motor, select Performance Requirements from the Main Menu and select Orbital Maneuvers. You will be presented with the Input Menu shown in Fig. A.2. Note that any three of four entries are adequate. Type 1.831 on the Velocity Change line and press Enter. (Velocity change is in metric units because mission design computations are commonly done in metric units.) Type 290 on the Specific Impulse line and press Enter. Note that either Initial or final weight is required, not both. Skip a line using the Down Arrow, or Tab Key; type 1500 on the Final Weight line. Before you press Enter, the screen should look like Fig. A.2. After you press Enter, PRO will calculate weights and display the results shown in Fig. A.3. The upper-stage motor must carry 1355.6 lb of usable propellant. At ignition, the upper stage will weigh 2855.6 lb. Weights are calculated in any system you use for the initial weights entry. If you had entered the final weight as 680.4 kg, the propellant weight computed by PRO would be 614.9 kg. If you had entered the final weight in stones (107.1), you would get propellant weight in stones (96.8).

The last line in the results screen asks if a printed report is desired. A yes response (just press Enter) will print results to a wide variety of printers. You will asked to enter the heading you want; this input is optional. Two lines are provided. Type N or n to indicate that no report is desired.

One-Axis Maneuver

A spacecraft makes attitude changes for such purposes as a turn for a star fix or a turn to fix the instruments on a target or a turn to point the propulsion system. These attitudes changes are performed by roll maneuvers about each of the spacecraft axes. PRO provides the ability to calculate the propellant consumed

```
                          ORBITAL MANEUVERS


                Given:
                Velocity Change, km/s = 1.831
                Specific Impulse, s   = 290
                Final Weight          = 1500

                Results:
                Propellant Weight, lb = 1355.6339
                Initial Weight, lb    = 2855.6339


                Do You Want Printed Copy? Y/N Y
```

Fig. A.3   Propellant required for orbital maneuvers.

in such maneuvers by calculating one axis at a time. Note that these maneuvers are usually performed in pairs; a turn to target and a turn back to cruise attitude.

To analyze one axis of such a maneuver, go to One-Axis Maneuver from the Propulsion Requirements Menu. The Input Menu is shown in Fig. A.4. For example, calculate the propellant required to rotate a spacecraft 7 deg around the x axis, given:

Moment of inertia about the x axis = 850 slug-ft^2
Number of thrusters to be fired = Usually 2
Thrust each = 2 lb

```
                         ONE AXIS MANEUVER


               Moment of Inertia, slug-ft²  =   850
               Angle of Rotation, deg       =   7
               Number of Thrusters          =   2
               Thrust Each, lb              =   2
               Specific Impulse, s          =   113
               Moment Arm, ft               =   7

               Results:
               Minimum Time Maneuver, s        = 3.851662
                        Propellant Used, lb    = 0.136342
               Minimum Propellant Maneuver, s = 185.461237
                        Propellant Used, lb    = 0.001416


               Do You Want Printed Copy? Y/N Y



     Esc = Back Up    Move = ↑↓ PgUp PgDn    Edit = ←    Accept = Enter
```

Fig. A.4   One-axis maneuver Input Menu.

Specific impulse = 113 s (obtained from the Pulsing Performance Menu assuming a long engine off time)

Moment arm = 7 ft (distance from the center of mass to the centerline of the thruster. It is difficult to get a radius more than 7 ft in a launch vehicle shroud.

After these entries are made, the results will be displayed as shown in Fig. A.4. Two results are presented: 1) one for a minimum time maneuver in which the thrusters burn continuously and 2) one for a minimum propellant consumption maneuver in which a minimum impulse bit of 20-ms pulse width is used.

One-Sided Limit Cycle

A one-sided limit cycle can occur when the spacecraft is subjected to an external torque such as solar torque or a gravity gradient torque. To calculate the propellant consumed in such a cycle as a function of mission time, go to One-sided Limit Cycle from the Propulsion Requirements Menu. You will get an Input Menu as shown in Fig. A.5. As an example, enter the following parameters:

Moment of inertia = 850 slug-ft^2

External torque = 3 ft-lb

Number of thrusters = 2

Thrust each = 0.5 lb

Specific impulse = 113 (from Pulsing Performance function-long engine off time)

Moment arm = 7 ft (distance from the center of mass to the centerline of the thruster)

Pulse width = 0.02 s (time from ON signal to OFF signal)

If the control torque is less than the external torque, you will get an error message. Figure A.5 shows the screen after the last entry is made. Propellant is being

```
                        ONE SIDED LIMIT CYCLE



                Moment of Inertia, slug-ft²  =   850
                External Torque, ft-lb       =   3
                Number of Thrusters          =   2
                Thrust Each, lb              =   0.5
                Specific Impulse, s          =   113
                Moment Arm, ft               =   7
                Pulse Width, s               =   0.02

                Results:
                Propellant Used, lb/s = 0.00379267 lb/s
                Minimum Control Band  = ±0.00002752 deg
                     (To avoid forced limit cycle)

                Do You Want Printed Copy? Y/N Y



     Esc = Back Up    Move = ↑↓ PgUp PgDn    Edit = ←   Accept = Enter
```

Fig. A.5   One-sided limit-cycle menu.

consumed at the average rate of 0.00379267 lb/s of elapsed mission time. This may not seem like much, but the propellant consumed in a day would be

$$0.00379267 \text{ lb/s} \times 86400 \text{ s/day} = 327.7 \text{ lb/day}$$

The minimum control band shown in Fig. A.5 refers to the control band at which forced limit-cycle operation would start, with resulting sharp increase in propellant consumption.

Try the same entries again, with external torque increased to 8 ft-lb. You will get an error message because the control torque is smaller than the external torque. Hence, the vehicle is uncontrollable.

Spin-Axis Precession

Spinning spacecraft are maneuvered by precessing the spin axis. PRO provides for calculating the propellant consumed in this activity. Selecting Spin-Axis Precession from the Main Menu displays the Input Menu shown in Fig. A.6. As an example, input the following:

Angle of precision = 3 deg
Rate of rotation = 1 rpm (spacecraft rate of rotation about the spin axis.)
Moment of inertia = 600 slug-ft^2 (about the spin axis)
Number of thrusters = 1
Thrust each = 0.2 lb
Specific impulse = 113 s (from Pulsing Performance Menu, assuming a long engine off time)
Moment arm = 4 ft (from the center of mass to the thruster centerline)

The results of this calculation are shown in Fig. A.6. The pulse width should be large enough to be possible in a real valve, that is to say, greater than about 0.015 s.

```
┌──────────────────────────────────────────────────────────────────┐
│                      SPIN AXIS PRECESSION                          │
├════════════════════════════════════════════════════════════════════┤
│                                                                    │
│                                                                    │
│            Angle of Precession, deg    =  3                        │
│            Rate of Rotation, rpm       =  1                        │
│            Moment of Inertia, slug-ft² =  600                      │
│            Number of Thrusters         =  1                        │
│            Thrust Each, lb             =  0.2                      │
│            Specific Impulse, ·s        =  113                      │
│            Moment Arm, ft              =  4                        │
│                                                                    │
│            Results:                                                │
│            Propellant Used, lb = 0.007278                          │
│            Pulse Width, s      = 0.036392                          │
│                                                                    │
│                                                                    │
│            Do You Want Printed Copy? Y/N Y                         │
│                                                                    │
│                                                                    │
│   Esc = Back Up    Move = ↑↓ PgUp PgDn    Edit = ←    Accept = Enter │
└──────────────────────────────────────────────────────────────────┘
```

Fig. A.6   Spin-axis precession Input Menu.

```
                    REACTION WHEEL UNLOADING
```

```
              Wheel Momentum, lb-ft-s =   20
              Number of Thrusters     =   2
              Thrust Each, lb          =   .2
              Specific Impulse, s      =   150
              Moment Arm, ft           =   7

              Results:
              Propellant Used, lb = 0.019048
              Maneuver Time, s    = 7.142857

              Do You Want Printed Copy? Y/N  Y
```

```
  Esc = Back Up     Move = ↑↓ PgUp PgDn     Edit = ←     Accept = Enter
```

Fig. A.7   Unloading Magellan reaction wheel.

The pulse width must be small compared to the period of one rotation; holding the thruster ON for a full rotation results in zero precession of the spin axis.

Reaction-Wheel Unloading

Many systems use reaction wheels to control attitude. In this case, propulsion is needed to unload the wheel periodically. Select Reaction-Wheel Unloading from the Propulsion Requirements Menu.

The Magellan reaction-wheel maximum momentum was 20 lb-ft/s. It was unloaded using a 0.2-lb thruster pair that delivered a specific impulse of 150 s at a moment arm of 7 ft. It took a 7.14 second burn to unload the wheel, and 0.01905 lb of propellant were consumed; see Fig. A.7.

A.5   Rocket Engine Design

The Rocket Engine Design function, accessed from the Main Menu, provides an estimate of engine performance by applying empirical efficiencies to theoretical frozen equilibrium specific impulse, C^* and C_f. Estimates of engine dimensions and weight are made using least-squares curves fits of actual engine data. This sort of information is often called *rubber engine data*; it is useful in early studies before real engine data are available. The performance information, I_{sp}, C_f, and C^*, is accurate to about 3%; the dimension and weight data are not based in physics and are susceptible to designer's choice and greater variance.

Select Rocket Engine Design from the Main Menu. Next, select the Propellant, hydrazine, nitrogen tetroxide/MMH, solid, cold gas, or other, and the Input Menu, shown in Fig. A.8, will appear. The default values of molecular weight, ratio of specific heat, and combustion temperature are displayed for editing along with

```
                  INPUT ROCKET ENGINE DESIGN DATA
────────────────────────────────────────────────────────────────

          Propellant                    = N2H4
          Molecular Weight              = 13.04
          Ratio of Specific Heats       = 1.27
          Combustion Temperature, °R    = 2210
          Cf Efficiency                 = 98
          C* Efficiency                 = 95
          Thrust, lb                    = 100
          Area Ratio                    = 50
          Chamber Pressure, psia        = 350




    Esc = Back Up    Move = ↑↓ PgUp PgDn    Edit = ←    Accept = Enter
```

Fig. A.8   Rocket engine design.

C_f and C^* efficiencies. These default values can be changed now or through the SETUP program. As an example, input:
Propellant = N_2H_4
Molecular weight = 13.04
Ratio of specific heats = 1.27
Combustion temperature = 2210°R
C_f efficiency = 98
C^* efficiency = 95
Thrust = 100 lb
Area ratio = 50
Chamber pressure = 350 psia
After the last entry, the calculation results will be displayed; see Fig. A.9.

A.6 Pulsing Engine Performance

Pulsing an engine has a profound effect on its performance because the combustion gases do not reach steady-state temperature. As a result, specific impulse can be substantially reduced. PRO estimates pulsing specific impulse from a statistical fit of parametric data taken by Olin Aerospace Company (then Rocket Research).[15]

After selecting Pulsing Engine Performance from the Main Menu, you will be asked to select the Propellant, hydrazine, nitrogen tetroxide/MMH, or other. The Input Menu shown in Fig. A.10 will appear. The default values for the combustion gases of the selected propellants will be displayed. They may be changed temporarily from this menu or changed globally from the SETUP program. For an example, make the following inputs:
Propellant = Hydrazine
Steady-State I_{sp} = 233 s
Molecular weight = 13.04

```
                 ESTIMATED ROCKET ENGINE PERFORMANCE

         Results:
         Specific Impulse, lbf-s/lbms     = 232
         Characteristic Velocity, fps     = 4166
         Vacuum Thrust Coefficient        = 1.7937
         Throat Area, Hot, sq in          = 0.1593
         Throat Diameter, Hot, in         = 0.4503
         Exit Area, Hot, sq in            = 7.9645
         Exit Diameter, Hot, in           = 3.1845
         Propellant Flow Rate, lbm/sec    = 0.4305
         Valve Inlet Pressure, psia       = 462.0000

         Dimensions:
         Catalyst Bed Diameter, in        = 3.8
         Motor Length, in                 = 7.6
         Motor Weight, lb                 = 4.4

         Do You Want Printed Copy? Y/N Y
```

Fig. A.9   Estimated rocket engine performance.

Ratio of specific heats = 1.27
Area ratio = 50
Prefire thruster temp = 200°F. (It is common practice to heat hydrazine thrusters to raise pulsing I_{sp}.)
Pulse width = 0.020 s (If you enter a pulse width smaller than 0.010 s, you will get an error message. A pulse width this small is not currently practical, and it is out of the database this function is based on. If you decide to proceed, an answer will be produced but it may not be reliable.)
Engine off time = 3000 s (The longer the engine is off because pulses, the colder the thruster will be at ignition.)

```
                        PULSING PERFORMANCE


               Propellant                 =  HYDRAZINE
               Steady State Isp, s         =  233
               Molecular Weight           =  13.04
               Ratio of Specific Heats    =  1.27
               Area Ratio                 =  50
               Pre-Fire Thruster Temp, °F =  200
               Pulse Width, s             =  .020
               Engine Off Time, s         =  3000

               Results:
               Pulsing Isp, s = 126.8027

               Do You Want Printed Copy? Y/N Y




Esc = Back Up   .Move = ↑↓ PgUp PgDn    Edit = ←   Accept = Enter
```

Fig. A.10   Pulsing Performance.

The calculated pulsing I_{sp} is 126.8 s, or 54% of steady state. If you repeat the calculation for several duty cycles and temperatures, you can see how the I_{sp} is affected. Some values are shown in Table A.1. Figure A.11 compares the results produced by PRO with test data compiled by Olin Aerospace Company.[15] The database underlying the pulsing I_{sp} function is entirely from monopropellant hydrazine firings.

A.7 Blowdown System Performance

The selection of a blowdown system significantly simplified the hardware required for a system. A blowdown system is more complex to analyze because propellant tank pressure is time-variant and depends on a number of factors, including temperature and propellant consumed. The PRO software implements the equations developed in the text to produce time-dependent performance.

The blowdown calculation makes use of system files containing a set of propulsion parameters. After selecting Blowdown Performance from the Main Menu, you will be asked to select:

> 1 Load a System File
> 2 Edit a System File
> 3 Create a System File

If you have not built a file, select 3, and a series of menus will help you collect physical property data; the system file must then be completed as shown in Fig. A.13. You should enter a tank volume that includes all the space available to the propellant and the pressurizing gas, that is, the volume of the tank when pressurized plus the volume of the feed lines to the engine valve minus the volume of the diaphragm or bladder. The pressurizing gas and propellant weights should be the full prelaunch load. The flow coefficient is obtained from the pressure drop from tank pressure to chamber pressure.

If you already have a system file or if you want to use the Magellan 100-lb engine system, select Load a System File; the System File Name window, shown in Fig. A.14 will appear.

To load a file, simply type the file name, *without the extension*, and press Enter. File names can be in upper or lower case. If the file is in the current directory, a

Table A.1 Pulsing Specific Impulse

Engine Off, s	Pulse Width, s	Temp °F	I_{sp}, s
ss	ss	2210	233
3000	0.02	200	127
3000	0.02	60	113
3	0.02	200	146
3	1.00	60	233

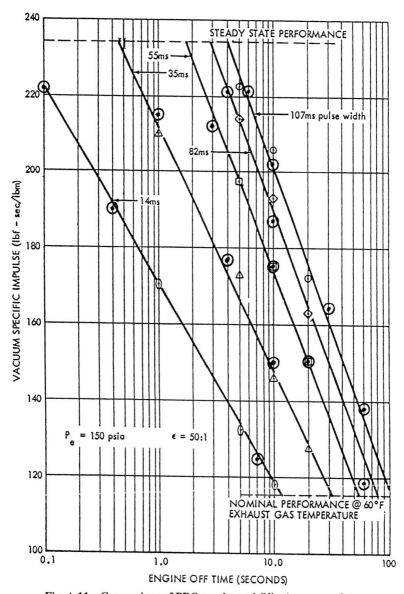

Fig. A.11   Comparison of PRO results and Olin Aerospace data.

```
                    PROPULSION SYSTEM FILE
━━━━━━━━━━━━━━━━━━━━━━━━━━━━━━━━━━━━━━━━━━━━━━━━━━━━━━━━━━━━━━━━━━

            System Name                 =
            Pressurization System        =   BLOWDOWN
            Pressurant Gas               =   HELIUM
            Gas Constant                 =   386.25
            Gas Loaded Weight, lb        =
            Propellant                   =   HYDRAZINE
            Prop. Density, lb/ft3        =   62.67
            Molecular Weight ,           =   13.04
            Combustion Temperature, °R   =   2210
            Specific Heat Ratio          =   1.27
            Propellant Loaded, lb        =
            Tank Volume, cu ft           =
            Flow Coefficient             =
            Steady State Isp, lb-s/lb    =
            Area Ratio                   =
            Throat Area, sq in           =
            Thrust Coefficient           =
            Characteristic Velocity      =


      Save & Continue = Alt-S   Continue = Alt-C    Back Up = Esc
```

Fig. A.12   Blowdown performance input.

drive and path are unnecessary. All the system files in the current directory are shown in the directory portion of the window.

The Input Menu is shown in Fig. A.12. For example, call up the MGN100 system file and input the following:

Number of thrusters firing = 2
Propellant remaining = 285.12 lb
Propellant temperature = 60°F
Prefire chamber temp. = 180°F
Press Down Arrow or Enter to skip tank pressure.
Gas temperature = 60°F

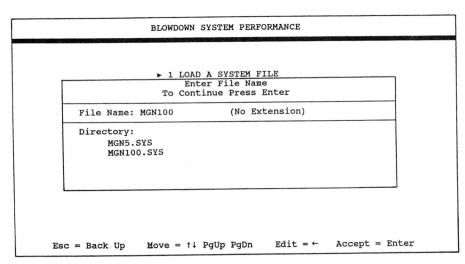

Fig. A.13   Parameters required for a system file.

```
                    BLOWDOWN SYSTEM PERFORMANCE


              System Name              =   MGN 100# ENGINE
              Number of Thrusters Firing  =   2
              Propellant Remaining, lb    =   285.12
              Propellant Temperature, °F  =   60
              Pre-Fire Chamber Temp, °F   =   180
              Tank Pressure, psia         =
              Gas Temperature, °F         =   60

                   Input gas temperature
                   OR tank pressure.





    Esc = Back Up    Move = ↑↓ PgUp PgDn    Edit = ←    Accept = Enter
```

Fig. A.14   Entering a file name.

A window will appear asking if performance is to be computed in pulse mode
or in continuous firing. Select Pulse Mode, and a second window will appear to
accept input describing the pulse duty cycle and the number of pulses to calculate.
If you elect to print results, a tabulated report will be printed and a portion of
the results will be displayed on screen as shown in Fig. A.15. Had you selected
Continuous Operation, an Input screen would have appeared to accept Burn time,
Calculation Increment, and Print Increment. The PRO calculations assume that the
ullage gas expansion is isentropic during a continuous burn and isothermal during
pulsing.


```
               Propulsion System Performance-Results


         Conditions at time = 0:
              Tank Pressure           = 379.03 psia
              Ullage Volume           = 2.1012 cu ft
              Propellant Flow Rate    = 1.0933 lb/s
              Chamber Pressure        = 179.81 psia
              Thrust                  = 76.53 lb
              Pulsing Isp,            = 140 s
              Pw                      ± 0.0250 s
              Engine Off Time         = 5 s

                    Pt, psia   Wp, lb/s   Pc, psia    F, lb
      Pulse  1      378.95     1.0932     179.79      76.52
      Pulse  2      378.87     1.0930     179.77      76.51
      Pulse  3      378.79     1.0929     179.74      76.50
                 Press any key to continue...
```

Fig. A.15   Blowdown system performance pulsing.

```
                    Propulsion Weight Statement

          Component              Unit Wt    No.    Total Wt

      PROPELLANT: Useable        283.5       1       284
            Unuseable              9.7        1        10
            Tank                  41.9        1        42
            Control Device         3.3        1         3
            Valves                 1.3       13        17
            Lines & Fittings       9.5        1        10
      PRESSURANT:      Gas         0.9        1         1
            Tank                   3.3        1         3
            Valves                 0.3        5         2
            Lines & Fittings       4.3        1         4
      MISC HARDWARE                2.0        1         2
      MOTOR:    100#               5.1        8        41
      MOTOR:      5#               0.8        8         6
      MOTOR:     .2#               0.4       12         5

      Total                                           430


  Enter = Next Field    Alt-C= Continue    Alt-I = Ins Line    Alt-P = Print
  Esc   = Main Menu     PgDn = Bot Line    Alt-D = Del Line    Alt-S = Save
```

Fig. A.16   Propulsion weight statement.

A.8   System Weight Statement

When you select function 5, System Weight Statement, from the Main Menu, you will be asked to input a file name from an input screen similar to Fig. A.14. All the system weight files in the current directory will be shown on the directory list. MGNPROP.WTS is a practice file you can use to get acquainted with the PRO features. After you enter a file name (no extension), the Propulsion weight Statement screen will appear as shown in Fig. A.16. You may make or change entries in any row or column on the spreadsheet. When the cursor moves away from an entry by use of the arrow keys, Enter, or Tab, PRO considers the entry acceptable and recalculates based on it. The following control keys are active on the Propulsion Weight Statement screen:

Enter, Arrow Keys, Tab = Accept an entry and move to the next field. The weight statement is recalculated.

Esc = Return to the Main Menu (without saving the weight statement).

Alt-C = Continue; return to the Main Menu without saving.

PgUp = Moves to the top line in a column.

PgDn = Moves to the bottom line in a column.

Alt-I = Inserts a new line at the location of the cursor.

Alt-D = Deletes the line at the location of the cursor.

Alt-P = Print a copy of the weight statement and a sequenced weight statement.

Alt-S = Saves a copy of the weight statement.

Alt-S means hold down the Alt key and simultaneously press the upper case or lower case *S*.

A.9   Utilities

Selecting Utilities from the Main Menu transfers you to the Utilities Menu for a selection. Figure A.17 shows the utilities provided by PRO. To select a utility,

```
                         UTILITIES


            ▶ 1 THEORETICAL PERFORMANCE
              2 EQUATION OF STATE
              3 TANK DESIGN
              4 ZERO-G CONTROL DEVICES
              5 CONVERSIONS







 Esc = Back Up    Move = ↑↓ PgUp PgDn    Edit = ←   Accept = Enter
```

Fig. A.17   Utilities Menu.

enter the menu number (1–5), or use the arrow keys to highlight your selection and press Enter. Each of the utility functions is discussed in the following sections.

Theoretical Performance

The theoretical performance function uses the ideal rocket thermodynamic relations described in the text to calculate theoretical, vacuum, frozen equilibrium performance. The propellant is first selected, which defines the thermodynamic properties of the exhaust gas. The theoretical performance Input Menu, Fig. A.18, is then displayed.

```
                  THEORETICAL PERFORMANCE
                Vacuum; Frozen Equilibrium


         Given:
         Propellant                  = HYDRAZINE
         Molecular Weight            = 13.04
         Ratio of Specific Heats     = 1.27
         Combustion Temperature, °R  = 2210
         Area Ratio                  = 50

         Results:
         Isp, s      = 249.4995
         C*, fps     = 4385.8983
         Cf          = 1.8303
         Pc/Pe       = 845.1255
         Ve, fps     = 7767.9146

         Do You Want Printed Copy? Y/N Y
```

Fig. A.18   Results of theoretical performance calculation.

You can see from Fig. A.18 that the thermodynamic properties can be changed if desired and that either the area ratio or pressure ratio of the engine must be entered. After the pressure ratio or area ratio is entered, the theoretical performance of an ideal engine is calculated. If area ratio is given, pressure ratio is calculated, and vice versa.

Equation of State

This function solves the equation of state to define the properties of a gas. Given any three properties of the following four—temperature, pressure, volume, or weight—the remaining property is calculated. After selecting Equation of State from the Utilities Menu, you will be asked to select a gas from the following list:
Helium
Nitrogen
Air
MMH vapor
N_2H_4 vapor
Other

If you want to use a gas not listed, select Other; then enter the gas name and specific gas constant. After the gas is selected, the Input Menu, Fig. A.19, will be displayed. As an example, try the Mariner C pressurant sphere, which held 1.000 lb of nitrogen. What was the volume of the sphere if the pressure was 3000 psia when the gas temperature was 70°F? Enter the three known parameters, and PRO will calculate a value of 0.067673 ft^3.

Tank Design—Mariner C Pressurant Tank

A large percentage of the dry weight of a propulsion system is in tankage. Typically, these tanks are spherical titanium tanks although tanks with barrel sections are sometimes used. The Tank Design function provides for rapid estimation of tank weights. Select Tank Design from the Utilities Menu, and you will be asked

```
                        EQUATION OF STATE



                Given:
                Gas              = NITROGEN
                Temperature, °F  = 70
                Pressure, psia   = 3000
                Gas Weight, lb   = 1.0000

                Results:
                Volume, cu ft    = 0.067673

                Do You Want Printed Copy? Y/N  Y
```

Fig. A.19   Equation of State Menu.

```
                    SPHERICAL TANK DESIGN

        Given:
        Material                 = TITANIUM
        Allowable Stress, psi    = 96000
        Material Density, lb/cu in = .16004
        Thickness Tolerance, in  = .002
        Internal Volume,  cu ft  = 0.067673
        Maximum Pressure, psia   = 3300
        Results:
        Inside Diameter, in      = 6.0672
        Max Membrane Thickness, in = 0.0541
        Membrane Weight          = 1.0200
        Girth Weld Land Weight, lb = 0.6724
        Penetration Weight, lb   = 0.3403
        Structural Attachment, lb = 0.0407
        Total Tank Weight, lb    = 2.0733

        Do You Want Printed Copy? Y/N Y
```

Fig. A.20   Mariner C pressurant sphere.

to select either a spherical tank or a barrel tank. The Input Menu, Fig. A.20 for spherical tanks, will be displayed.

Titanium, the default material, is shown along with representative properties. (If you wish to use a material other than titanium, enter the name, allowable stress, and density.) When you enter a volume and maximum pressure, the properties of the tank are displayed. For example, the Mariner C pressurant tank had an allowable stress of 96,000 psi, a volume of 0.067673 ft^3, and a maximum pressure of 3300 psia. Enter these parameters, omitting commas and leading zeros, and PRO will calculate estimated size and weight for the Mariner C tank; the results are shown in Fig. A.20. PRO estimates the weights of the land area provided for the hemispherical weld, the penetrations, and the structural attachments as well as the eight of the tank membrane.

The actual weight of the Mariner C tank was 1.90 lb[23] vs a calculated value of 2.07 lb.

Zero-g Control Devices—Mariner C Bladder

The most common types of zero-G propellant control devices are bladders, diaphragm, and capillary devices. PRO provides a function for estimating the weight of two of these types. Select Zero-G Control Devices from the Utilities Menu and you will be asked to select either a bladder or a diaphragm; then you will be asked to select the material see Fig. A.21. The name and density of the selected material will be displayed. If you wish to use a material that is not offered, enter the name and density. Then, input the volume of propellant to be contained and the bladder thickness. For example, the Mariner C bladder contained 0.35064 ft^3 of propellant. Estimating a thickness of 0.075 in. produces the following results:

Bladder weight,
Outside diameter,
Inside diameter,
Bladder volume, in.

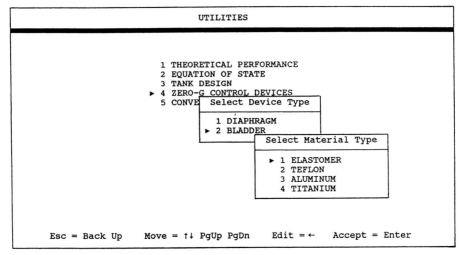

Fig. A.21   Zero-*g* Devices Selection Menu.

The propellant tank must accommodate the volume of the bladder as well as the volume of the propellant. The actual weight of the Mariner C bladder was 1.0 lb[23] vs a calculated weight of 0.948 lb.

Conversions

Selecting Conversions from the Utility Menu will display the eight types of conversion PRO can perform; see Fig. A.22. Select the conversion you want to make, for example 2, Force: Newtons to Pounds, and a window will appear to accept your input in newtons. Once you have entered force in newtons, the screen will display your input and the pounds force equivalent.

A.10   Sample Problems

In this section, six sample problems are described and worked. In each case, the problem is described, the needed PRO commands are listed, and the PRO results are included. The sample problems are:
1) Orbital Maneuver: Magellan Venus Orbit Insertion
2) Solid Rocket Motor Design: IUS Stage I
3) Liquid Engine Design: Magellan 100-lb Thruster
4) Pulsing Engine Performance: RRC 5-lb Thruster
5) Blowdown System Performance: Magellan Orbit Trim
6) Tank Design: Viking Orbiter Propellant Tank

Orbital Maneuver: Magellan Venus Orbit Insertion

The Magellan spacecraft was injected into Venus orbit on August 10, 1990. The injection energy was supplied by a Thiokol Corporation Star 48B. Thrust vector control was provided by aft-facing 100-lb monopropellant thrusters. The velocity change provided by the Star 48B was 2.715 km/s after compensation was made for the contribution of the monopropellant thrusters. The Magellan mass at ignition

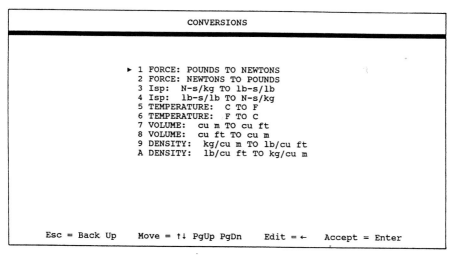

Fig. A.22   Conversions Menu.

was 3285 kg; the mass expelled by the Star 48B was 2024 kg. The effective specific impulse delivered by the solid motor cannot be measured in flight but it can be calculated by PRO.

The necessary PRO commands are as follows:

1) Choose Propulsion Requirements from the Main Menu. (The easiest way is simply to press Enter.)

2) Choose Orbital Maneuvers from the Propulsion Requirements Menu. Press Enter. The Orbital Maneuvers Input Menu will now be displayed, showing:

Velocity change, km/s
Specific impulse, s
Initial weight
Final weight
You need to enter any three.

3) The cursor is on Velocity Change. Type 2.715 and press Enter.

4) Specific impulse is sought, and so press the Down Arrow once or press Enter again.

5) Type 3285. Press Enter. (Weights can be in any unit system.)

6) Type 1261 (3285-2024). Press Enter.

After you press Enter, the results screen will appear showing:

Results:
Specific impulse,
Propellant weight,
The effective specific impulse was 289.2 s.

Solid Rocket Motor Design: IUS Stage I

The IUS stage I solid rocket motor design is summarized in the *Inertial Upper Stage User's Guide*[26] and is an ideal check for the Rocket Design function in PRO. Use the following steps to see how well PRO can estimate the design of the motor.

1) Select Rocket Engine Design from the Main Menu.

2) Choose Solid from the Select Propellant Menu. A window will drop down, displaying the available solid propellants; use the Down Arrow to move to HTPB/AP/AL. Press Enter to select.

3) The Input Menu will be displayed showing the default properties of HTPB. Any of these properties may be changed; in addition, the following four motor parameters are needed:

Total impulse, lb/s

Average thrust, lb

Area ratio

Average chamber pressure

4) Use Enter of the Down Arrow to move down to Total Impulse. Type 6324882. Press Enter.

5) Type 41611 (average thrust in pounds). Press Enter.

6) Type 63.80 (area ratio). Press Enter.

7) Type 579 (average chamber pressure in psia). Before you press Enter, the screen will look like Fig. A.23.

After you press Enter, the results screen will be displayed.If you want a printed copy, press Enter. You will be given two lines to input a heading with a suggested first line. You can select one, two, or no heading lines. If there is a heading, the date will be added as the last line. Figure A.24 is the printed report for this calculation. If you do not want a report, type N, (upper or lower case) for no, and you will be returned to the Main Menu.

The PRO estimate and the actual motor parameters are listed in Table A.2. The engine performance estimates are based on theoretical performance and efficiencies normally achieved. The weight and dimensions are based on statistical curve fits of existing engines and are subject to wider variances.

```
INPUT ROCKET ENGINE DESIGN DATA



             Propellant                =  HTPB/AP/AL
             Molecular Weight          =  22
             Ratio of Specific Heats   =  1.26
             Combustion Temperature, °R =  6160
             Cf Efficiency             =  98
             C* Efficiency             =  93
             Density, lb/cu in         =  .0651
             Total Impulse, lb-s       =  6324882
             Average Thrust, lb        =  41611
             Area Ratio                =  63.80
             Avg Chamber Pressure, psia =  579




   Esc = Back Up     Move = ↑↓ PgUp PgDn     Edit = ←    Accept = Enter
```

Fig. A.23   IUS stage 1 motor design.

```
                  ESTIMATED ROCKET ENGINE PERFORMANCE
                              IUS STAGE I
                             03-29-1993
    Given:
          Propellant                  = HTPB/AP/AL
          Molecular Weight            = 22
          Ratio of Specific Heats     = 1.26
          Combustion Temperature, °R  = 6160
          Cf Efficiency               = 98
          C* Efficiency               = 93
          Density, lb/cu in           = .0651
          Total Impulse, lb-s         = 6324882
          Average Thrust, lb          = 41611
          Area Ratio                  = 63.80
          Avg Chamber Pressure, psia  = 579

    Theoretical Performance (Vacuum, Frozen Equilibrium):
          Specific Impulse, lbf-s/lbm  = 325
          Characteristic Velocity, fps = 5653
          Vacuum Thrust Coefficient    = 1.8541
          Pressure Ratio, Pc/Pe        = 1167
          Exit Plane Pressure, psia    = 0.4958
          Exhaust Velocity, fps        = 10173

    Estimated Real Engine Performance (Vacuum, Steady State):
          Specific Impulse, lbf-s/lbms = 296
          Characteristic Velocity, fps = 5257
          Vacuum Thrust Coefficient    = 1.8171
          Throat Area, Hot, sq in      = 39.5511
          Throat Diameter, Hot, in     = 7.0963
          Exit Area, Hot, sq in        = 2523.3622
          Exit Diameter, Hot, in       = 56.6820
          Burn Time, s                 = 152.0002
          Propellant Weight, lb        = 21301.3079

    Dimensions Assume Spherical Motor & Submerged Nozzle:
          Motor Diameter, in           = 87.7
          Motor Length, in             = 157.1
          Motor Weight, lb             = 22541.1
```

Fig. A.24   PRO report: IUS stage 1 design.

Liquid Engine Design: Magellan 100-lb Engine

The 100-lb monopropellant thruster flown on Magellan and DMSP was designed and built by Olin Aerospace Company. To compare the actual engine performance with the PRO estimate:

1) Select Rocket Engine Design from the Main Menu.

2) Chose Hydrazine from the Select Propellant Menu.

3) The Input Menu will be displayed showing the default properties of hydrazine. Any of these properties may be changed; in addition, the following four parameters are needed:

Thrust, lb
Area ratio
Chamber pressure, psia

Table A.2   Comparison of IUS Stage I Design and PRO Estimate

Parameter	PRO	Actual
Specific impulse, s	296	295.5
Throat diameter, in.	7.1	6.5
Exit diameter, in.	56.7	51.8
Burning time, s	152	152
Propellant weight, lb	21,301	21,404
Motor weight, lb	22,541	22,981
Motor diameter, in.	87.7	92.0
Motor length, in.	157.1	123.98

4) Use Enter or the Down Arrow to move down to Thrust. Type 110.8. Press Enter.

5) Type 50 (area ratio). Press Enter.

6) Type 258 (chamber pressure in psia). Before you press Enter, the screen will look like Fig. A.25.

After you press Enter, the results screen will be displayed. If you want a printed copy, press Enter. You will be given two lines to input a heading with a suggested first line. You can select one, two, or no heading lines. If there is a heading, the date will be added as the last line. Figure A.26 shows the results of this calculation. If you do not want a report, type N (upper or lower case) for no, and you will be returned to the Main Menu. The PRO estimate and the actual motor parameters are listed in Table A.3 The engine performance estimates are based on theoretical performance and efficiencies normally achieved. The weight and dimensions

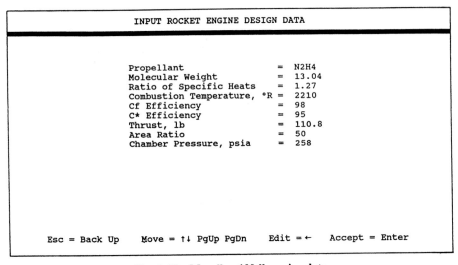

```
                    INPUT ROCKET ENGINE DESIGN DATA



              Propellant                =  N2H4
              Molecular Weight          =  13.04
              Ratio of Specific Heats   =  1.27
              Combustion Temperature, °R =  2210
              Cf Efficiency             =  98
              C* Efficiency             =  95
              Thrust, lb                =  110.8
              Area Ratio                =  50
              Chamber Pressure, psia    =  258




    Esc = Back Up    Move = ↑↓ PgUp PgDn    Edit = ←    Accept = Enter
```

Fig. A.25   Magellan 100-lb engine data.

```
                ESTIMATED ROCKET ENGINE PERFORMANCE
                MAGELLAN 100 POUND QUAL ENGINE
                         03-29-1993

    Given:
          Propellant              = N2H4
          Molecular Weight        = 13.04
          Ratio of Specific Heats = 1.27
          Combustion Temperature, °R = 2210
          Cf Efficiency           = 98
          C* Efficiency           = 95
          Thrust, lb              = 110.8
          Area Ratio              = 50
          Chamber Pressure, psia  = 258

    Theoretical Performance (Vacuum, Frozen Equilibrium):
          Specific Impulse, lbf-s/lbm  = 249
          Characteristic Velocity, fps = 4385
          Vacuum Thrust Coefficient    = 1.8303
          Pressure Ratio, Pc/Pe        = 845
          Exit Plane Pressure, psia    = 0.3052
          Exhaust Velocity, fps        = 7767

    Estimated Real Engine Performance (Vacuum, Steady State):
          Specific Impulse, lbf-s/lbms = 232
          Characteristic Velocity, fps = 4166
          Vacuum Thrust Coefficient    = 1.7937
          Throat Area, Hot, sq in      = 0.2394
          Throat Diameter, Hot, in     = 0.5521
          Exit Area, Hot, sq in        = 11.9715
          Exit Diameter, Hot, in       = 3.9042
          Propellant Flow Rate, lbm/sec = 0.4770
          Valve Inlet Pressure, psia   = 351.6000

    Dimensions:
          Catalyst Bed Diameter, in  = 4.0
          Motor Length, in           = 8.9
          Motor Weight, lb           = 4.7
```

Fig. A.26   PRO report: Magellan 100-lb engine.

are based on statistical curve fits of existing engines and are subject to wider variances.

Pulsing Engine Performance: 5-lb Engine

Pulse mode operation substantially reduces the specific impulse of an engine because of energy losses heating the engine. The PRO function that estimates pulsing engine performance is based on Olin Aerospace Company testing of a 2-lb engine at various duty cycles. Duty cycles are characterized by pulse width, the time between valve open and valve closed commands, and engine off time. Let us compare the PRO pulsing estimates with test data from five Magellan 22N (5-lb) engines.

1) Select Pulsing Engine Performance from the Main Menu.
2) Select Hydrazine from the Propellant Choice Menu.
3) Move cursor down one line. Type 232.5 (steady-state I_{sp}). Press Enter.

Table A.3   Comparison of 100-lb Qual Engine
and PRO Estimate

Parameter	PRO	Actual
Specific impulse, s	232	233.5
Thrust coefficient	1.793	1.676
Propellant flow rate, lb/s	0.477	0.475
Valve inlet pressure, psi	351	350
Motor weight, lb	4.7	5.1

4) Move down to Area Ratio. Type 60 and press Enter.

5) At Prefire Thruster Temp, type 200 (°F) and press Enter.

6) Move down to Pulse Width. Type .020 and press Enter.

7) Move down to Engine off time. Type 20(s) and press Enter.

The projected pulsing specific impulse is 127.5 s. Repeating the process for initial thruster temperatures of 400, 600, 800 and 1000°F produces the specific impulse predictions of 145.6, 161.6, 176.2, and 189.7, respectively. Figure A.27 shows the projections by PRO compared to test results[47] on five Magellan 5-lb thrusters. The predictions made by PRO are within the scatter and toward the low side of engine test results.

Blowdown System Performance: Magellan Orbital Trim

In this section, we will use PRO to estimate the performance of the Magellan propulsion system during the first Venus orbit trim maneuver. The PRO estimate will be compared with telemetered data from the actual maneuver.

1) Select Blowdown System Performance from the Main Menu.

2) You will be asked if you want to:

> 1 Load a System File
>
> 2 Edit a System File
>
> 3 Create a System File

Select Load a System File.

3) The File Name Input window will be displayed. Note the directory of file choices. Type MGN100 and press Enter. (The maneuver was performed using the Magellan 100-lb engine system. The detail data on the system are contained in the system file named MGN100. You can use the Edit System File function to review the contents of the file if you wish.)

4) Once the file is retrieved, PRO will select the needed information and display the Input Menu.

5) Move down to Number of Thrusters Firing and type 4. Press Enter.

6) Enter:

Propellant remaining, lb = 285.12

Propellant temperature, °F = 68

Prefire chamber temp., °F = 200

Tank Pressure, psia = 380

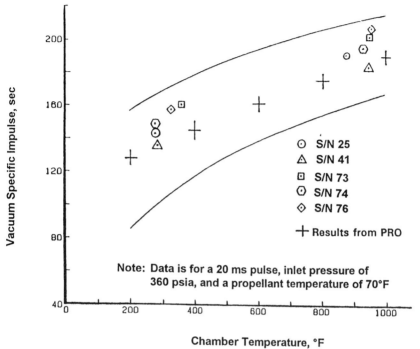

Chamber Temperature, °F

Fig. A.27   Pulsing specific impulse: PRO results vs test results.

7) You will then be asked if you want:
1 Performance in Continuous Firing
2 Performance in Pulse Mode
Select 1, Performance in Continuous Firing.
8) You will be asked to input calculation parameters;
input:
Burn time, s = 5.5
Calculation increment, s = .5
Print increment, s = 1
9) The conditions at time = 0 will be displayed and you will be asked if you want a printed copy. Enter Y. (It is almost essential to copy performance data to printer; only a summary can be put on screen.)
Figure A.28 compares the PRO calculations with the telemetered data. The PRO data are within about 1% of the actual results. Note that the telemetered data are slightly delayed and that telemetered pressure resolution is about 2 psia. PRO does not predict the post shutdown pressures.
The report generated by PRO is shown in Fig. A.29.

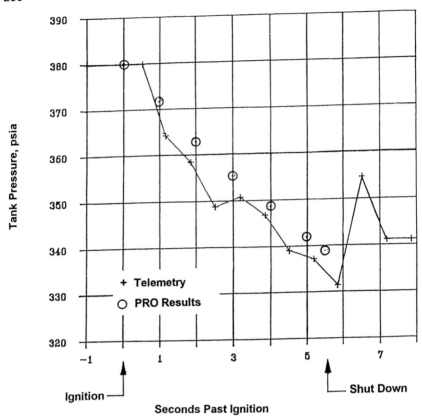

Fig. A.28 Magellan orbit trim vs PRO results.

Tank Design: Viking Orbiter Barrel Tank

The Viking Orbiter carried two identical large titanium barrel tanks for the bipropellant system used for Mars orbit insertion. The tanks were 35 in. in diameter and 25.3299 ft^3 in volume.[32] To compare the actual characteristics of the Viking tank with PRO tank design calculation, use the following steps:

1) Select Utilities from the Main Menu.
2) Select Tank Design from the Utilities Menu.
3) You will be asked to select spherical or barrel tanks; select Barrel.
4) The Barrel Tank Input Menu will be displayed, with the default properties of titanium displayed.
5) Move down one line and change the Allowable Stress to 125000 psi.
6) Move down to Internal Volume, type 25.3299 (ft^3), and press Enter.
7) Enter Max Tank Pressure = 890 psia.
8) Enter inside diameter = 35.00 in.

The Input Menu is shown in Fig. A.30. The results of PRO design calculations are shown in Fig. A.31 and compared to the actual Viking Orbiter tank in Table A.4.

```
                    PROPULSION SYSTEM PERFORMANCE
                         MGN 100# ENGINE
                          03-09-1995

Given:
     Number of Thrusters Firing  = 4
     Propellant Remaining, lb     = 285.12
     Propellant Temperature, °F   = 68
     Pre-Fire Chamber Temp, °F    = 200
     Tank Pressure, psia          = 380

Conditions at Time = 0
     Gas Temperature            = 62.15 °F
     Ullage Volume              = 2.1045 cu ft
     Propellant Density         = 62.9752 lb/cu ft
    'Propellant Flow Rate       = 1.7961 lb/s
     Chamber Pressure           = 245.60 psia
     Thrust Per Engine          = 104.53 lb


Time Since Ignition = 1 s
     Tank Pressure              = 371.61 psia
     Gas Temperature            = 57.54 °F
     Ullage Volume              = 2.1330 cu ft
     Propellant Flow Rate       = 1.7667 lb/s
     Propellant Remaining       = 283.32 lb
     Chamber Pressure           = 241.57 psia
     Thrust Per Engine          = 102.82 lb

Time Since Ignition = 2 s
     Tank Pressure              = 363.64 psia
     Gas Temperature            = 53.11 °F
     Ullage Volume              = 2.1611 cu ft
     Propellant Flow Rate       = 1.7385 lb/s
     Propellant Remaining       = 281.56 lb
     Chamber Pressure           = 237.72 psia
     Thrust Per Engine          = 101.18 lb

Time Since Ignition = 3 s
     Tank Pressure              = 356.06 psia
     Gas Temperature            = 48.83 °F
     Ullage Volume              = 2.1887 cu ft
     Propellant Flow Rate       = 1.7115 lb/s
     Propellant Remaining       = 279.82 lb
     Chamber Pressure           = 234.03 psia
     Thrust Per Engine          = 99.61 lb

Time Since Ignition = 4 s
     Tank Pressure              = 348.85 psia
     Gas Temperature            = 44.71 °F
     Ullage Volume              = 2.2159 cu ft
     Propellant Flow Rate       = 1.6855 lb/s
     Propellant Remaining       = 278.11 lb
     Chamber Pressure           = 230.48 psia
     Thrust Per Engine          = 98.10 lb

Time Since Ignition = 5 s
     Tank Pressure              = 341.97 psia
     Gas Temperature            = 40.73 °F
     Ullage Volume              = 2.2426 cu ft
     Propellant Flow Rate       = 1.6606 lb/s
     Propellant Remaining       = 276.42 lb
     Chamber Pressure           = 227.07 psia
     Thrust Per Engine          = 96.65 lb
```

Fig. A.29   PRO report: reconstruction of Magellan OTM-1.

Table A.4   Comparison of Viking Orbiter
Barrel Tank and PRO Estimate

Parameter	PRO	Actual
Tank weight, lb	111.4	110.9
Length, inside, in.	57.16	57.16

```
                    BARREL TANK DESIGN



        Material                  =  TITANIUM
        Allowable Stress, psi     =  125000
        Material Density, lb/cu in =  .16004
        Thickness Tolerance, in   =  .002
        Internal Volume,  cu ft   =  25.3299
        Maximum Pressure, psia    =  890
        Tank Inside Diameter, in  =  35






   Esc = Back Up    Move = ↑↓ PgUp PgDn    Edit = ←   Accept = Enter
```

Fig. A.30   Barrel Tank Input Menu: Viking Orbiter parameters.

```
                    BARREL TANK DESIGN


        Results:
        Tank Length (Inside), in      = 57.1604
        Barrel Length, in             = 22.1604
        Max Barrel Thickness, in      = 0.1286
        Barrel Weight, lb             = 50.3335
        Max Hemisphere Thickness, in  = 0.0643
        Sphere Weight, lb             = 39.7485
        Girth Weld Land Weight, lb    = 18.2372
        Penetration Weight, lb        = 0.4041
        Structural Attachment, lb     = 2.1745
        Total Tank Weight, lb         = 110.8977

        Do You Want Printed Copy? Y/N Y
```

Fig. A.31   Barrel tank results: Viking Orbiter tank.

A.11   Hard Drive Installation

To copy PRO to a hard disk, go to the root directory and create a directory for PRO using the MKDIR command. For example:

MD PROP

Then, go to the new directory:

CD PROP

Put the PRO disk in drive A. To copy the disk to the directory you have created, enter:

COPY A:*.*

DOS will copy all the PRO files to your hard disk; the files are:
PRO.EXE—the main PRO code file
GASSES.DAT—a file containing the properties of various gases
MATERIAL.DAT—a file containing material properties
MGN100.SYS—a typical system file for practice, containing the performance parameters from the Magellan 100-lb pound monopropellant engine system
MGN5.SYS—a typical system file for practice, containing the performance parameters from the Magellan 5-lb monopropellant engine system
MGNPROP.WTS—a typical propulsion system weight statement
PROPEL.DAT—a file containing fluid propellant properties
SETPROP.EXE—support software that provides review and edit the SETUP and .DAT files
SETUP.PRO—a file containing default parameters
SOLID.DAT—a file containing properties of various solid propellant types.
All these files must be in the same directory for PRO to operate properly.

To start: Go to the hard disk directory where you installed PRO, or put the PRO disk in drive A.

Type PRO (caps or lower case), and press ENTER

When the title screen appears, PRO is ready to use.

Equipment Requirements

IBM PC, XT, AT, PS2, or compatible*
240K RAM Memory
One disk drive
Printer highly desirable
Math coprocessor (PRO will operate with or without one)

---

*IBM is a registered trademark of International Business Machines Corporation.

Appendix B
Propulsion Design Data

Appendix B is the professional reference appendix. After you understand the basics of propulsion design, this is where you look up all those numbers you are going to need. The default parameters used in PRO are tabulated here and can be changed using the SETUP program.

B.1 Conversion Factors

Table B.1 Conversion Factors (Data in part, from Ref. 45, pp. 689–698)

	Multiply	by ···	to Get	
Angles	deg	57.2957795131	rad	
Area	ft^2	0.09290304	m^2	
Density	$lb/in.^3$	27679.905	kg/m^3	
	lb/ft^3	16.018463	kg/m^3	
	lb/ft^3	0.01601846	g/cc	
Force	lbf	4.448221615	N	
Impulse	lb/s	4.448221615	N-s	
I_{sp}	lbf/s-lbm	9.8066516	N-s/kg	
Length	ft	0.3048	m	Exact
	in.	0.0254	m	Exact
Mass	lbm	0.45359237	kg	Exact
Momentum	lbf-ft/s	1.355818	N-m/s	
Pressure	psi	6894.7572	N/m^2	
	psf	47.880258	N/m^2	
	psf	4.4882429	kg/m^2	
Speed	km/s	3280.84	ft/s	
	rpm	0.1047198	rad/s	
Torque	ft-lb	1.355818	N-m	
Volume	ft^3	0.0283168	m^3	
Moment of Inertia	$slug-ft^2$	1.355914	kg/m^2	
Velocity	ft/s	0.3048	m/s	Exact
	km/s	3280.84	ft/s	

Table B.2   Constants

Gravitational Constant	32.1740 ft/s^2
Density of water (4°C)	62.4266 lb/ft^3
R_{N_2}	55.16
R_{He}	386.3

B.2 Propellant Properties

Table B.3   Properties of Propellant

Propellant	Symbol	Molecular Weight	Freezing Point °F	Boiling Point °F	Density @68°F	Vapor Pressure psia	°F
Chlorine trifluoride	ClF_3	92.46	−105.4	53.15	1.825	20.8	110
Fluorine	F_2	38	−363	−307	1.51[a]	5.0	−322
Hydrazine	N_2H_4	32.05	35.6	236.3	1.008	0.2	68
Hydrogen	H_2	2.02	−435	−423	0.071[a]	1.02	−435
MMH	$CH_3N_2H_3$	46.08	−62.1	188.2	0.8765	0.70	68
Nitric Acid	HNO_3	63.02	−42.9	185.5	1.513	0.93	68
Nitrogen Tetroxide	N_2O_4	92.02	11.8	70.1	1.447	13.92	68
Oxygen	O_2	32	−361.8	−297.6	1.14[a]	7.35	−308
RP-1	$CH_{1.9-2.0}$	175	−48	362 to 500	0.806	0.02	68
UDMH	$(CH_3)_2N_2H_2$	60.10	−71	146	0.793	2.38	68

[a]At normal boiling point.

Table B.4   Properties of Anhydrous Hydrazine (from Ref. 15, pp. 13, 14)

Property	English	Metric
Molecular weight	32.04	
Boiling point	236.3°F	113.5°
Critical properties	P_c = 145 atm	
	T_c = 716°F	T_c = 380°C
		d_c = 0.231 g/cc
Dielectric constant	51.7 @ 77°F	51.7 @ 25°C
Explosive limits in air	4.7% lower	
	100% upper	
Flash point (open cup)	125.6°F	52°C
Freezing point	35.6°F	2°C
Heat capacity @77°F	0.737 Btu/lb	23.62 cal/mole
Surface tension @ 77°F	0.004568 lb/ft	66.67 dyne/cm

Table B.5   Density of Hydrazine (from Ref. 21)

Temp. °F	Density lb/ft^3	Temp. °C	Density kg/m^3
32	64.0399	0	1025.82
40	63.7966	10	1017.03
50	63.4912	20	1008.15
60	63.1840	25	1003.68
70	62.8751	30	999.19
80	62.5644	40	990.13
90	62.2520	50	980.98
100	61.9378	60	971.74
110	61.6192		
120	61.3043		
130	60.9849		
140	60.6638		

$\rho = 65.0010 - 0.0298(°F) - 8.7023E - 6(°F)^2$, lb/ft^3.
$\rho = 1025.817 - 0.8742(°C) - 0.0005(°C)^2$, kg/m^3.

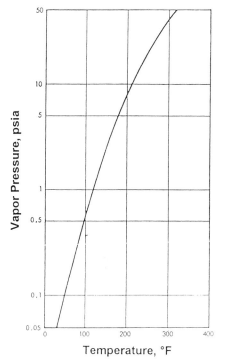

Fig. B.1   Vapor pressure of anhydrous hydrazine (from Ref. 15, p. 20).

Table B.6   Properties of Nitrogen Tetroxide (data from Ref. 46)

Property	English	Metric
Molecular weight	92.02	
Boiling point @ 1 atm	70.1°F	21.15°C
Critical properties, T_c	316.4°F	158°C
P_c	1470 atm	
ρ_c		0.56 g/cc
Density, liquid @ 68°F		1.447 g/cc
Density, gas @ 70°F, 1 atm		3.40 g/l
Freezing point @ 1 atm	11.7°F	−11.3°C
Specific heat, gas, Cp, 27°–67°C, 1 atm		1.62 cal/g-°C
Surface tension @ 25°C		26.5 dyne/cm
Thermal conductivity @ 55°C	0.021 BTU/hr-ft²-°F	
Vapor pressure @ 70°F	14.7 psia	

Density of Nitrogen Tetroxide: (From 60° to 160°F)

$$\rho = 91.060 - 0.0909(T - 60) \ \text{lb/ft}^3$$

where $T = $ liquid temperature °F.

Table B.7   Specific Gravity of Common Propellants

Propellant	Specific Gravity	Temperature, °F
Hydrazine	1.008	67
	0.984	120
Liquid hydrogen	0.071	−423
Monomethylhydrazine	0.8788	67
	0.8627	100
Nitrogen tetroxide	1.447	67
	1.37	120
Liquid oxygen	1.23	−320
	1.14	−297
RP-1	0.807	76
	0.58	300

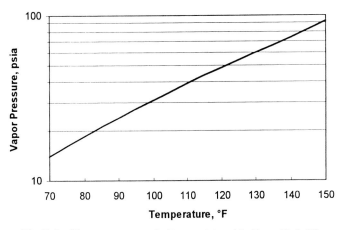

Fig. B.2   Vapor pressure of nitrogen tetroxide (from Ref. 46).

B.3 PRO: Files

The following tables display the design data stored in the support files on the PRO software disk. You can change the information in any of these files by using the SETPRO software on the PRO disk.

Table B.8   PRO: Liquid Propellant Property Files

Propellant	Property	Value
Hydrazine		
	Molecular weight	13.04
	Ratio of specific heats	1.27
	Combustion temperature, °R	2210
	Density, lb/ft^3	62.67
N$_2$O$_4$/MMH		
	Molecular weight	21.52
	Ratio of specific heats	1.25
	Combustion temperature, °R	6145
	Bulk density, lb/ft^3	71.67
	Mixture ratio	1.50

Table B.9   PRO: Gas Property Files

Gas	Property	Value
Helium		
	Molecular weight	4
	Ratio of specific heats	1.659
Nitrogen		
	Molecular weight	28
	Ratio of specific heats	1.4
Air		
	Molecular weight	28.9
	Ratio of specific heat	1.4
Freon-14		
	Molecular weight	88.0
	Ratio of specific heats	1.3

Table B.10   PRO: Material Property Files

Material	Property	Value
Elastomer		
	Density, $lb/in.^3$	0.036
Teflon		
	Density, $lb/in.^3$	0.077
Aluminum		
	Density, $lb/in.^3$	0.098
Titanium		
	Density, $lb/in.^3$	0.016
	Allowable stress psi	100,000

Table B.11   PRO: Solid Propellant Files

Propellant	Property	Value
PBAA/AP/AL		
	Molecular weight	22
	Ratio of specific heats	1.26
	Combustion Temperature, °R	6260
	Density, lb/in.^{3}	0.064
	Burning rate, in./s	0.32
	Pressure exponent	0.35
PBAN/AP/AL		
	Molecular weight	22
	Ratio of specific heats	1.26
	Combustion temperature, °R	6260
	Density, lb/in.^{3}	0.064
	Burning rate, in./s	0.55
	Pressure exponent	0.33
CTPB/AP/AL		
	Molecular weight	22
	Ratio of specific heats	1.26
	Combustion temperature, °R	6160
	Density, lb/in.^{3}	0.064
	Burning rate, in./s	0.45
	Pressure exponent	0.40
HTPB/AP/AL		
	Molecular weight	22
	Ratio of specific heats	1.26
	Combustion temperature, °R	6160
	Density, lb/in.^{3}	0.065
	Burning rate, in./s	0.282
	Pressure exponent	0.30
PS/AP/AL		
	Molecular weight	22
	Ratio of specific heats	1.67
	Combustion temperature, °R	5460
	Density, lb/in.^{3}	0.062
	Burning rate, in./s	0.31
	Pressure exponent	0.33
PVC/AP/AL		
	Molecular weight	22
	Ratio of specific heats	1.26
	Combustion temperature, °R	6260
	Density, lb/in.^{3}	0.064
	Burning rate, in./s	0.45
	Pressure exponent	0.35

B.4   Planetary Data

Table B.12   Gravitational Parameters, Mean
Equatorial Radii (from Ref. 10, p. 173)

Planet	μ km^3/s^2	R_0 km
Mercury	22032.1	2439
Venus	324858.8	6052
Earth	*398600.4*	*6378.14*
Mars	42828.3	3397.4
Jupiter	126711995.4	71492.4
Saturn	37939519.7	60268.4
Uranus	5780158.5	25559.4
Neptune	6871307.8	25269.10
Pluto	1020.9	1500
Moon	4902.8	1738
Sun	132712439935.5	696000

Appendix C
Glossary

action time, t_a Burning time of a solid rocket grain plus the time to burn the sliver; usually taken as the time from 10% thrust at ignition to 10% thrust at burnout.

area ratio, e The ratio of exit area to throat area in a converging-diverging nozzle; always larger than one.

binders The complex hydrocarbons that serve as solid fuels and have as a secondary function providing mechanical strength, binding all the grain ingredients.

bipropellants A combination of rocket propellants always including an oxidizer and a fuel. For example, liquid oxygen and liquid hydrogen are a bipropellant combination.

burning, rate r The linear velocity at which a cigarettelike strand of solid propellant will burn as measured by the velocity of the flame front, usually in inches per second.

burning rate exponent, n Burning rate, in a solid propellant, increases proportionately with the absolute pressure raised to the burning rate exponent.

burning time, t_b Time of vigorous combustion of a solid grain; usually taken as the time from 10% thrust during ignition to 90% thrust during burnout (also called *burn time*).

chamber pressure, P_c The total pressure in a rocket engine measured before the converging nozzle.

characteristic exhaust velocity, $C*$ A figure of merit for a propellant combination; high is good.

combustion chamber The chamber, in a rocket engine, in which the propellants are ignited and burn.

composite propellants Any solid rocket propellant that is a mechanical mixture of oxidizers, fuels, and additives. Composite propellants are the industry standard.

critical pressure ratio The pressure ratio across a nozzle that will result in sonic flow through the nozzle.

cryogenics Propellants that are gases at normal temperatures but are used in propulsion systems chilled below their boiling points, for example, liquid oxygen. Also the science dealing with very cold liquids.

cumulative damage A property exhibited by solid motor grains. The allowable stress in the grain decreases with cumulative time under stress.

deflagration limit The minimum chamber pressure at which combustion will continue.

double-based propellants A solid motor propellant composed of more than one explosive, each of which acts as oxidizer and fuel.

213

effective exhaust velocity   An exhaust velocity that has been adjusted to allow the small pressure area term of a vacuum engine to be neglected.

effective propellant weight   The loaded weight less the burnout weight of a solid rocket motor (may be more or less than the propellant weight).

erosive burning   Burning, in a solid motor, that erodes the surface of the grain as well as consuming it chemically.

expansion ratio   Same as area ratio.

frozen equilibrium   As assumed condition of fixed chemical composition of the gaseous products of combustion as they flow through the rocket nozzle; as opposed to shifting equilibrium.

geysering   A phenomenon observed in cryogenic liquids that causes destructive eruption of propellant in long feed lines; caused by heat transfer into the propellant which, in turn, causes rapid vaporization.

hybrid propellants   A propellant combination consisting of a liquid and a solid propellant.

impulse, I   The product of thrust and time; the area under a thrust-time curve; also the product of total propellant weight and specific impulse.

liner   The purpose of the liner in a solid rocket is to extinguish the flame and to insulate the case.

LOX   An acronym for liquid oxygen.

minimum impulse bit, I_{min}   The impulse generated by an engine/valve combination, with the minimum time between on-signal and off-signal.

mixture ratio, MR   The weight ratio of oxidizer to fuel consumed in a bipropellant system.

molecular weight, M   The relative weight of molecules of gases compared to oxygen, which is arbitrarily set at 32 units.

monopropellants   Any one of a number of unstable chemicals that will decompose, under suitable conditions, releasing energy and producing hot gas that can be used in a rocket engine.

neutral burning   A solid motor grain that produces constant, or almost constant, thrust.

optimum area ratio   The area ratio that produces an exit plane pressure equal to the local atmospheric pressure.

outage   The mass of liquid propellant that is not usable because of the difference between the mixture ratio of propellant actually loaded and the mixture ratio actually burned; applicable to bipropellant systems only; can be driven to very nearly zero by active control of the mixture ratio burned.

progressive burning   A solid motor grain producing thrust that increases with time.

propellant holdup   The same as unusable propellant.

RP-1   An acronym for rocket propellant 1, a hydrocarbon fuel similar to kerosene.

solid propellants   A rocket propellant combination in which both propellants are solids at normal temperatures.

specific gas constant, R   The proportional constant that relates pressure volume and temperature for a unit weight of a specific gas, $PV/T = R$ and $R = 1545/M$.

specific impulse, I_{sp} The thrust generated per unit mass flow rate of propellant through an engine; used with any type of rocket propulsion system; usually expressed in seconds or seconds-pound force per pound mass.

shifting equilibrium An assumed condition of changing chemical composition in the exhaust gases as they flow through a rocket nozzle.

shutdown impulse The total impulse delivered by a propulsion system after the closing signal is sent to the thrust chamber valves.

sliver The solid propellant residual left unburned after the web extinguishes.

sliver fraction The fraction of the loaded propellant weight that is left unburned when the web extinguishes.

throat area, A_t The minimum area in a converging-diverging nozzle.

thrust coefficient, C_f A figure of merit for a nozzle when a given gas (ratio of specific heats) flows through it.

total impulse, I The total impulse delivered by a propulsion system is the total propellant mass consumed times the average specific impulse. It is also equal to the average thrust times the burn time.

UDMH An acronym for unsymmetrical dimethyl hydrazine, a storable rocket fuel.

ullage The volume of the gas space above the propellant in a propellant tank; most frequently expressed as a percentage of the propellant volume (an expression borrowed from the wine industry).

unusable propellant The total mass of propellant that cannot be burned for all reasons including outage, trapped, drop-out, loading errors, and vapor loss.

volumetric loading fraction In a solid motor, the ratio of the grain volume to the case volume, excluding the nozzle.

volumetric mixture ratio The volume ratio of oxidizer to fuel used in a propulsion system; also the ratio of fuel density to oxidizer density times the mixture ratio.

web The maximum radial thickness of a solid propellant grain.

web fraction, b_f The ratio of the web thickness, of a solid grain to the radius of the grain.

web thickness The minimum dimension from the port surface to the liner interface measured radially (a solid motor term).

References

[1]Ridpath, I., *Space,* Hamlyn, New Work, NY, 1981.

[2]Osman, T., *Space History,* St. Martins, New York, NY, 1983.

[3]Corliss, William R., *Scientific Satellites,* NASP SP-133, NASA Washington, DC, 1967.

[4]Cochran, C. D., et al. (eds.), *Space Handbook,* AU-18, Air University Press, Maxwell AFB, AL, 1985.

[5]Koelle, H. H. (ed.), *Handbook of Astronautical Engineering,* McGraw-Hill, New York, 1961.

[6]Ring, E., and Brown, C. D. et al. (eds.), *Rocket Propellant and Pressurization Systems,* Prentice-Hall, Englewood Cliffs, NJ, 1964.

[7]Sutton, G. P., *Rocket Propulsion Elements,* Wiley, New York, 1986.

[8]Zucrow, M. J., *Principles of Jet Propulsion,* Wiley, New York, 1952.

[9]Seifert, H. S., (ed.), *Space Technology,* Wiley, New York, 1959.

[10]Brown, C. D., *Spacecraft Mission Design,* AIAA Education Series, AIAA, Washington, DC, 1992.

[11]Hohmann, Walter, *Erreicharkeit der Himmelskörper,* Munich, Germany, 1925.

[12]Agrawal, B. N., *Design of Geosynchronous Spacecraft,* Prentice-Hall, Englewood Cliffs, NJ, 1986.

[13]Isakowitz, Steven, J., *International Reference Guide to Space Launch Systems,* AIAA, Washington, DC, 1991.

[14]Wertz, J. R., and Larson, W. J. (eds.), *Space Mission Analysis and Design,* Kluwer Academic Press, Dordrecht, The Netherlands, 1991.

[15] *Monopropellant Hydrazine Design Data,* Rocket Research Corporation, Seattle, WA.

[16]Schartz, W. T., Cannova, R. D., Cowley, R. T., and Evans, D. D., "Development and Flight Experience of the Voyager Propulsion System," AIAA Paper 79-1334, 1979.

[17]Dressler, G. A., Morningstar, R. E., Sackheim, R. L., and Fritz, D. E., *Qualification of Augmented Electrothermal Hydrazine Thruster,* AIAA, New York, 1981.

[18]Smith, W. W., Smith, R. D., Yano, S. E., Davies, K., and Lichtin, D., "Low Power Hydrazine Arojet Flight Qualification," AIDAA/AIAA/DGLR/JSASS, 22nd International Electric Propulsion Conference (Viareggio, Italy), 1991.

[19]Morningstar, R. E., Kaloust, A. H., and Macklis, H., "An Operational Satellite Propulsion System Providing for Vernier Velocity, High and Low Level Attitude Control and Spin Trim," AIAA Paper 72-1130, 1972.

[20]Sackheim, R. L., "Survey of Space Applications of Monopropellant Hydrazine Propulsion Systems," Tenth International Symposium on Space Technology and Science, (Tokyo, Japan), AIAA, New York, 1973, (AIAA Paper 74-42367).

[21]Bell, Tim, "Magellan OTMI Propulsion Report," (rev. A), Martin Marietta, Denver, CO, February 1992.

[22]Hearn, H. C., "Design and Development of a Large Bipropellant Blowdown Propulsion System," AIAA Paper 93-2118, AIAA, Washington, DC, 1993.

[23]Schmitz, B. W., Groudle, T. A., and Kelley, J. H., "Development of the Post Injection System for Mariner C Spacecraft," JPL Technical Report No. 32-830, JPL, 1966.

[24] *LANDSAT 3 Reference Manual,* General Electric Space Division, Philadelphia, PA, 1978.

[25]Holmberg, N. A., Faust, R. P., and Holt, H. M., *Viking '75 Spacecraft Design and Test Summary*, NASA Reference Publication 1027, NASA, Washington, DC, 1980.

[26]*The Inertial Upper Stage Users Guide*, D290-11011-1, Boeing Aerospace Co., Seattle, WA, 1984.

[27]Hughes Aircraft Company, *Pioneer Venus Case Study in Spacecraft Design*, AIAA, New York, 1979.

[28]Rausch, R. J., Johnson, J. T., and Baer, W., *Intelsat V Spacecraft Design Summary*, AIAA, New York, 1981.

[29]Hamlin, K., McGrath, D. K., and Lara, M. R., *Venus Orbit Insertion of the Magellan Spacecraft Using a Thiokol STAR 48B Rocket Motor*, AIAA Paper 91-1853, 1991.

[30]Frazier, R. E., *HEAO Case Study in Spacecraft Design*, TRW Report 26000-100–102, AIAA, New York, 1981.

[31]TRW Systems Group, *Thermal Network Modeling Handbook*, N75-30483, NASA, Washington, DC.

[32]Schmit, D. D., Anderson, J. W., and Vote, F. C., "Long-Life Bipropellant System Demonstration, Viking Orbiter Propulsion System," *Journal of Spacecraft and Rockets*, Vol. 18, No. 4, 1981.

[33]Bohnhoff, K., "Retro Propulsion Module (RPM), Bipropellant Unified System for Gallileo, 33rd International Astronautical Federation Congress," AIAA, New York, (IAF Paper 82-326).

[34]Virdee, L. S., Chang, C. P., and Dest., L., *In-Orbit Performance of Interlsat VI Liquid Propulsion System*, AIAA Paper 93-2515, 1993.

[35]Schmit, D. D., Leeds, M., and Vote, F., "In Flight Performance of the Viking 75 Orbiter Propulsion System," AIAA Paper 77-894, 1977.

[36]Barber, T. J., Krug, F. A., and Froidevaux, B. M., "Initial Galileo Propulsion System In-Flight Characterization," AIAA, Paper 93-2117, 1993.

[37]Cole, T. W., Frisbee, R. H., and Yavrouian, A. H., "Analysis of Flow Decay Potential on Galileo," AIAA Paper 87-2016, 1991.

[38]Dest, L. R., Bouchez, J. P., Serafini, V. R., Schavietello, M., and Volkert, K. J., "INTELSAT VI Spacecraft Bus Design," *COMSAT Technical Review*, Vol. 21, Spring 1991.

[39] *Solid Rocket Motor Performance Analysis and Prediction*, NASA SP 8039, Washington, DC, June 1971.

[40]Northam, G. B., and Lucy, M. H., "On the Effects of Acceleration Upon Solid Rocket Performance," AIAA Paper 68-530, 1968.

[41]Rogers, W. P., "Report of the Presidential Commission on the Space Shuttle Challenger Accident," GPO, Washington, DC, 1986.

[42]Sola, F. L., "Development of the Explorer Solid Rocket Motor," AIAA Paper 89-2954, 1989.

[43]Greer, H., and Griep, D. J., "Dynamic Performance of Low Thrust, Cold Gas Reaction Jets in a Vacuum," *Journal of Spacecraft and Rockets*, Vol. 4, No. 8, 1967, pp. 983–990.

[44]Hall, S. E., Lewis, Mark J., and Akin, D. L., "Design of a High Density Cold Gas Attitude Control System," AIAA Paper 93-2583, 1993.

[45]Fogiel, M., *Handbook of Mathematical, Scientific and Engineering Formulas, Tables, Functions, Graphs, Transforms*, Research and Education Association, Piscataway, NJ, 1991.

[46] *Matheson Gas Data Book*, Matheson Co., East Rutherford, NJ, 1966.

[47]Frank, R. E., *Magellan Flight System Parameters Report*, VRM-SE-020-075, Martin Marietta, Denver, CO, July 1988.

Index

7.25